THE NATURE OF PHYSICS

THE NATURE OF PHYSICS

A PHYSICIST'S VIEWS ON THE HISTORY AND

PHILOSOPHY OF HIS SCIENCE

ROBERT BRUCE LINDSAY

BROWN UNIVERSITY PRESS PROVIDENCE, RHODE ISLAND

INTERNATIONAL STANDARD BOOK NUMBER: 0–87057–107–9

LIBRARY OF CONGRESS CATALOG CARD NUMBER: 68–10642

BROWN UNIVERSITY PRESS, PROVIDENCE, RHODE ISLAND 02912

PUBLISHED 1968. SECOND PRINTING 1971

PRINTED IN THE UNITED STATES OF AMERICA

DESIGNED BY MALCOLM GREAR DESIGNERS, INC.

TYPE SET IN BASKERVILLE BY CONNECTICUT PRINTERS, INC.

PRINTED BY CONNECTICUT PRINTERS, INC.

ON WARREN'S OLD STYLE

BOUND BY TAPLEY-RUTTER COMPANY, INC.

PREFACE

At various stages of their professional careers most physicists must come to grips with the philosophical and historical aspects of physics. In this book I have tried to present a synthesis of my own views about the meaning of physics as a science, views that have developed over the past forty years. Although they lean rather heavily on previously published articles, the contents of the first three chapters are not mere reprints of these; in every case they have been extensively revised, and in certain cases they represent evolutionary departure from earlier ideas.

The fourth chapter sets forth my general attitude toward the history of physics and illustrates it with excursions into certain specific problems: magnetism, atomism, the mechanics of falling bodies, and the history of acoustics, an old but still flourishing branch of science. This chapter owes its *raison d'être* to my conviction that the philosophy and history of physics, and indeed of science as a whole, are inseparable. This chapter is derived in large measure from previously published papers.

Grateful acknowledgment is made for the use of the following of my articles that form the basis of the book:

"Some Philosophical Aspects of Recent Atomic Theory," *Scientific Monthly*, XXVI (1928), 229–305.

"The Significance of Boundary Conditions in Physics," *Scientific Monthly*, XXIX (1929), 465–71.

"The Broad Point of View in Physics," *Scientific Monthly*, XXXIV (1932), 115–24.

"Causality in the Physical World," *Scientific Monthly*, XXXVII (1933), 330–37.

"The Meaning of Simplicity in Physics," *Philosophy of Science*, IV, No. 2 (1937), 151–67.

"A Critique of Operationalism in Physics," *Philosophy of Science*, IV, No. 4 (1937), 456–70.

"Measurement in Physics," *American Journal of Physics*, VIII (1940), 22–27.

"William Gilbert and Magnetism in 1600," *American Journal of Physics*, VIII (1940), 271–82.

"Physical Explanation and the Domain of Physical Experience," *Journal of the Washington Academy of Science*, XXXII (1942), 356–59.

"Galileo Galilei, 1564–1642, and the Motion of Falling Bodies," *American Journal of Physics*, X (1942), 285–92.

"On the Relation of Mathematics and Physics," *Scientific Monthly*, LIX (1944), 456–60.

"Pierre Gassendi and the Revival of Atomism in the Renaissance," *American Journal of Physics*, XIII (1945), 235–42.

"Operationalism in Physics Reassessed," *Scientific Monthly*, LXXIX (1954), 221–23.

"The Story of Acoustics," *Journal of the Acoustical Society of America*, XXXIX (1966), 629–44.

It is also appropriate to mention two books, the writing of which has helped to form my point of view: *Foundations of Physics*, written in collaboration with Henry Margenau (New York: John Wiley & Sons, Inc., 1936; New York: Dover Publications, 1957); and *The Role of Science in Civilization* (New York: Harper & Row, 1963).

I would like to acknowledge my debt to all those colleagues whose views, both in their published writings and in conversations, have helped to clarify my own ideas. It is plainly impossible to enumerate them all, but the Notes at the end of the book indicate to some extent material that has been particularly useful. Special mention in this connection must be made of Henry Margenau, F. S. C. Northrop, and Raymond J. Seeger.

I am also indebted to Miss Susan Desilets and Mrs. Ruth Wick, both of Brown University, for their help in typing the manuscript. It is a pleasure to acknowledge the valuable assistance rendered by Dr. Carol Schaefer and other members of the staff of the Brown University Press in the final preparation of the manuscript.

CONTENTS

THE NATURE OF PHYSICS

INTRODUCTION

In the chapters that follow physics will be considered as a branch of science—as a method for describing, creating, and understanding human experience. Human experience will be interpreted as including everything that happens to the human individual in every moment of his life, together with the reflections made on these happenings by his mind. Actually, of course, as a matter of convention physics has dealt with only a part of this vast experience, though it is a steadily increasing part, and no ultimate limit may legitimately be placed on it. Description here will mean the search for order or pattern in experience and the effort to describe it in the most meaningful way; it may also be called the endeavor to explain how things actually go, insofar as we can find regularity manifested in them. This process will lead to the development of the concept of physical law as a brief expression in appropriate symbolic language of a routine of physical experience.

Next to demand attention will be the creation of experience in physics —the performance of experiments, largely through the appropriate arrangement of physical objects and the performance of operations on them to see what happens. Stress will be laid on the idea that experimentation, often referred to as controlled sense perception, is more appropriately termed the creation of experience. Experimentation represents the manual side of a physicist's activity, though it is naturally accompanied by much mental activity as well, since it is not merely random manipulation. The creation of experience in physics goes on ceaselessly in countless laboratories throughout the world.

In addition to the description and creation of experience, the method of physics includes understanding. Here it will be regarded as the most important element in the process. Understanding encompasses the endeavor to discover why experience manifests the particular order or pattern exemplified by physical laws. Understanding results from the building of theories that are imaginative constructions of the mind, pictures of the world as it might be, and which, if it were, would lead to observed experience. A theory obviously demands the construction of symbolic concepts based on primitive notions like those of space and time and either suggested by experience or created out of whole cloth. A theory also involves setting up hypotheses or educated guesses as to relations connecting the constructs. From these hypotheses or postulates conclusions may be deduced, usually by the use of mathematical analysis. The hope of the theory builder is that such conclusions can be identified with observed physical laws. If this turns out to be so, the theory is thus far successful. But the theoretician demands even more: he insists that the theory predict a result with no counterpart in previous experience. If experimentation validates the prediction, the confidence of the physicist in the value of his theory is enormously increased. There are, of course, other important criteria for testing the success of physical theories, but prediction is by all odds the most decisive. All these elements in the method of physics will be treated in Chapter One.

Attention will be deliberately limited to the logical structure of physical theories, with a full realization of the very important psychological problems involved in the study of the actual processes of invention of theories by physicists. These problems constitute a field in which the returns so far have not been conspicuously successful.

The examination of the logical nature of physical theory itself poses many problems. These problems will be taken up in turn in Chapter Two, with particular emphasis on the nature of symbolism in physics, the origin of constructs and postulates, the role of law, and the appraisal of theories. Since the process of physical theorizing is essentially more one of invention than of discovery, we shall take a dim view of the use of the word truth in connection with physical theory, at least in the sense of a final definitive method of explanation that will hold good forever. The more modest and reasonable view that will emerge is that as new experience is continually created, new theories will be invented or old ones modified to meet the new demands. The word success will seem to be a more satisfactory description than truth in appraising the value of a physical theory. Hence the whole process of developing physics has a character more arbitrary than is commonly conceded by philosophers of science. Perhaps arbitrary is too strong a word, as it conveys a sense of capriciousness, and certainly the invention of physical theories is not a matter of caprice or faddism. Perhaps the words free invention are more appropriate than arbitrariness, since physicists believe that in inventing theories they have a right within certain broad limits to the free use of preference or taste. This has encouraged the use of the imagination untrammeled by undue adherence to the experience itself in the development of constructs and postulates. Who can doubt that it is this freedom that has led to such successful theories as quantum mechanics? It gives sanction to bold new approaches to a difficult domain when older methods fail or at any rate become cumbersome. Of course, the price of this freedom of invention is that the state of physical theory at any moment must be considered a tentative one. There is no finality to the process, and it must be expected that in all probability the physics of the future will be quite different from that of the present.

Throughout this book problems will be encountered that have long engaged the attention of professional philosophers, who have considered it their province to investigate such matters as the nature of experience and our knowledge of it, valid methods of thinking and expression, the logical character of the constructs used in science, and the ways in which scientists are permitted to use them without lapsing into inconsistency. These and other similar relations between physics and philosophy will prompt us to take a closer look in Chapter Three at some of the important problems in physics commonly called philosophical. In the first place particular attention will be paid to the nature of space and time as used in physics. These categories or modes of ordering experience are, to be sure, so fundamental in the interpretation of experience at any level, that no brief analysis can hope to do them justice.

After paying brief respects to the psychological significance of space and time to the human individual, the peculiar physical aspect of time will be noted as a mere parameter used for convenience in the equations of physics invented to describe change in physical systems. It will also be shown that measured values can be assigned to this parameter only by comparing the given physical system with another one—a clock—chosen as a standard. In this sense the use of time in physics reduces to spatialization. Just as in pure mathematics a convenient parameter in terms of which parametric equations are set up can be eliminated between the equations, yielding an equation independent of the parameter, so in the equations of physics containing the parameter t, t can often be eliminated, and the morphological equations that result are independent of time, describing form only. Such equations reflect the Parmenidean attitude toward experience as distinguished from the Heraclitean.

Actually, however, physics does not wholly encourage the elimination of time from theoretical analysis. It will therefore be necessary to examine the theoretical changes to which the time parameter t must be subjected when reference systems are altered. This leads directly to relativistic physics, to which some attention will be given. What relativity has done, of course, is to emphasize that if the concepts of space and time are to be used most effectively in theoretical physics, they cannot be treated as if they were independent. Rather they must be considered as inextricably connected. This in turn leads to the desirability of looking at physical phenomena as events taking place in a four-dimensional space-time continuum rather than as merely relations between objects located in space.

Such topics as the operational idea, the concepts of causality and determinism, and the relation between probability and statistics in physics will be examined next. No one questions the significance of the operational point of view, and this is recognized in the demand that all definitions of physical quantities that are to be identified with experience must contain operational (epistemic) as well as theoretical (constitutive) aspects in order to fulfill their purpose. On the other hand, to insist that all physical constructs must have operational meaning would seriously hamper the development of theoretical physics, for many constructs may have constitutive significance only and still be very useful. An example is the state function ψ in quantum mechanics.

Another important philosophical problem in physics is the meaning of the concept of causality. It is clear without much probing that the classical notions of cause and effect have little relevance to physics. Here causality will be taken as merely the exemplification of the existence of certain physical laws describing order in experience regardless of time. It will be found to be highly desirable to distinguish causality in this sense from determinism, with which it is often confused. The key idea in determinism is the ability to predict the future state of a physical system from its present state. In the case of complex systems with a multitude of components (such as a gas with many molecules) it is possible to set up statistical laws governing the behavior of the system and hence achieve causality, whereas determinism with respect to the behavior of the individual components is lost. Hence one can have causality without determinism. The situation is even more striking for single atoms, for which, according to the standard interpretation of quantum mechanics, indeterminism is fundamentally built into their behavior without regard to any statistical considerations. A look into this problem will lead to a discussion of the philosophical status of the so-called indeterminacy principle of quantum mechanics as it has developed in the history of twentieth-century physics. This will provide an opportunity to stress the intimate relation that often exists between the philosophy and history of physics.

Mention of the dynamics of systems of many components and the kinetic theory of gases immediately suggests probability and its role in physics. The concept is a very difficult one, and no mathematical analysis of it has ever been considered completely satisfactory. In some ways it may seem unfortunate that it has had to be employed in physics at all. Fortunately for the theoretical physicist he does not have to worry about it quite so much as the biological scientist, since almost his only use of it is in the calculation of average quantities identified with the results of measurement. Here he is entitled to experiment pretty much as he pleases with various types of probability distribution functions and to settle on those that prove the most convenient and yield results most nearly in agreement with experiment. In other words, he is not forced to provide a priori justification for the probability functions he employs. They enter conventionally into his calculations. This point will be emphasized by calling attention to J. Willard Gibbs's statistical mechanics and the lack of dependence of the averages obtained for systems of very large numbers of degrees of freedom on the particular probability density expressions employed.

The role of probability in physics extends also to experimentation, and in this connection measurement comes to mind, leading to a discussion of the theory of experimental errors. The essential arbitrariness of the procedures followed will be noted, with all due regard to their essential plausibility. But interest in measurement will center on the famous question of the interpretation of this process in terms of the object-instrument concept and the difficulties this has run into in connection with the "measurement" of purely constitutive quantities such as are encountered in the theory of atomic physics. After noting the popular presentation of the indeterminacy principle in terms of the object-instrument concept, it will be pointed out that this principle as a theorem in the standard theory of quantum mechanics has no necessary connection with this essentially macroscopic viewpoint. Niels Bohr's interest in the problem has kept the latter viewpoint very much alive and of course contributed to his feeling for the importance of his principle of complementarity. For Bohr the division of experience between observer and observed speaks directly for the great dichotomy in description between mutually exclusive modes. The notion is undoubtedly ingenious, but erected into a dogmatic principle it might well prove to be an obstacle to ultimate scientific advance.

The object-instrument and observer-observed dualism in science at once suggests consideration of the bearing on physics of the philosophical problem of realism versus idealism in the interpretation of experience. Though it is undoubtedly correct to say that no particular philosophical view has relevance to the success of a physicist, it is probably true that, by and large, experimental physicists tend toward philosophical realism in their epistemology, whereas theoretical physicists lean toward the idealistic standpoint. The idealistic epistemology of Sir Arthur Eddington will be examined as an extreme case of the latter. If his writings really mean what they appear to mean, Eddington evidently came to believe that the universe of man's experience is essentially the creation of man's mind and that if we could only understand precisely how man's mind works, we could derive the whole of physics—presumably all science as well—by purely theoretical methods, barring, of course, certain dimensional constants that are matters of accident, dependent on where one happens to be in the universe. Acting on this principle, Eddington proceeded to derive a number of dimensionless constants, like the ratio of the mass of the proton to that of the electron, that are fundamental in physics. He insisted that in this way the whole of our knowledge of physics could be developed without performing experiments. Few have been able to understand precisely how Eddington derived his results, but the suspicion is strong that his theoretical considerations do not differ logically from those of any other physical theory and that therefore his results, like other theoretical results, have no unique status but must stand or fall on their agreement or disagreement with experience, the creation of which goes on quite independently of Eddington's mode of thinking. All that is really unusual about his work appears to be the bold assertiveness of his idealistic philosophy. His work may indeed serve to stimulate further research into the psychology of invention in physics, and in this case it will have made a valuable contribution to science.

The search begun in Chapter Two for further criteria for appraising the success of a physical theory—in addition to the very powerful one of prediction of new experience—will lead in Chapter Three to deeper consideration of the criterion of simplicity. Naïvely speaking, it is an attractive yardstick, for most people think that they know when one set of ideas is simpler than another and prefer the former. The trouble is that closer inspection indicates that the idea may prove elusive just when it seems to have been pinned down. It turns out that simplicity means different things to different people and in the last analysis is apt to reduce to familiarity: familiar ideas seem simpler. An attempt will be made to see whether something can be made of the notion in terms of the purely quantitative measure of the number of so-called independent concepts entering into the structure of a theory. But it will be found impossible to establish the required independence in clear-cut fashion, and hence this measure will appear to fall to the ground. It is in some ways a pity that more cannot be made of what is superficially such an attractive notion, but the facts are against it.

Throughout this book it will be necessary to consider the symbolism of physics. For a long time it has been taken for granted that mathematics provides the most effective and indeed natural symbolism for use in physics. This came about not merely because of the quantitative aspects of physics. By using the shorthand symbolism of mathematics instead of the language of ordinary speech, the statements of physical hypotheses can be more succinctly expressed and can be manipulated more expeditiously and with less chance of logical error to produce deduced results. Hence mathematics pervades physics very thoroughly, and that this should be so is now taken for granted by all physicists and more and more by all scientists who have any occasion to use physics in their work. As a matter of fact, mathematics is rapidly coming to be the preferred language of all science.

The person who dislikes mathematics—and there are many—is apt to ask whether mathematics is really essential for the understanding of physics. If such a person is a layman who merely wants to know what physics is about, the answer is No, since anything that is expressed in the abstract symbolism of mathematics can be translated into the ordinary vernacular. In fact, this is the line taken by most popularizers of physics. But if deeper understanding is meant—the kind of understanding implied in the ability to use physics successfully in its application—then mathematics is inescapably the language of physics. This suggests the great importance of steadily creating new mathematics to match the creation of new physical experience. In the past, and in particular during the eighteenth and nineteenth centuries, this was actually done by the physicists themselves, as will be noted in the section on acoustics in Chapter Four. The scientists of that era who worked on the physics of sound laid the foundations for much of the mathematics that was refined by the pure mathematicians of the later nineteenth and early twentieth centuries. However, in more recent times it has become customary for physicists to find in the writings of pure mathematicians the material needed for their theoretical developments, and this process is accelerating in our own day. This suggests that the physicist and all others who use the language of mathematics should support the labors of pure mathematicians, since the language they are creating today will be in considerable measure the language of physics tomorrow—if history is any guide. This is important from another point of view, since the mathematics originated by professional mathematicians is apt to be good mathematics in the sense that its logical rigor is fairly well attested to. Much of the mathematics created by physicists themselves tends to be less logically rigorous, and this can lead to serious problems in testing theoretical results. If physicists are to use mathematics as a language, it must be language as free from logical inconsistencies as possible.

The fact that physics is the creation of people will be emphasized throughout this book. Physics therefore has a history, and this history plays a most significant role in explaining what physics is all about. Chapter Four therefore consists of observations on the history of physics and four excursions into special topics in its history. The history of science is a formidable discipline, demanding as it does all the professional qualifications of the historian and the scientist. It is fraught also with somewhat frustrating difficulties, not the least of which is the difficulty of interpreting ancient scientific records so as to grasp precisely the mental processes of the early investigators and the relations between those processes and the observations they report. Moreover, the history of science is of little value unless the meaning of early science is related to present-day science in the same topical area. Hence the history of science involves building theories about the early records much like the theories of science itself.

The aim of the four excursions in Chapter Four into the history of physics will not be to show the chronological evolution of physical ideas but to study certain views of great physicists in the domains of magnetism (Gilbert), atomic theory (Gassendi), mechanics (Galileo), and acoustics (Rayleigh). The first three excursions will emphasize certain problems—historical puzzles perhaps would be the right term for them—that one would like to solve in order to understand better the way in which the methodology of physical invention developed. In the case of Gilbert and magnetism, the puzzle is how he managed to develop a relation (albeit in geometrical form) between the magnetic dip and latitude on the earth's surface, although the analytical concept of a magnetic field was not known until more than two centuries after his time. Gassendi's work presents the mystery of how a seventeenth-century cleric was able to expound atomic theory at a time when it was anathema to many. In Galileo's case, the puzzle is why so much attention has been focused on his alleged discovery of the first law of motion and so little on what has proved to be a far more important invention—the way in which a deductive theory can be made to work successfully in physics, as exemplified by the deduction of the law of falling bodies.

The history of acoustics, in which Rayleigh figured as the encyclopedic summarizer, brings up many philosophical problems, not the least of which is the connection between sound phenomena and the development of the mathematics (partial differential equations) necessary to describe the behavior of continuous media. Wave propagation was first studied in order to understand acoustical phenomena, and this had an enormous influence on the trend of physical theorizing in other fields—for example, that of light. Acoustics offers a cogent illustration of a theory in which the fundamental concepts were initially as closely connected with raw experience as those of the mechanics of particles (displacement and velocity) but in which there was no hesitation in introducing much more abstract and imaginative concepts like acoustic impedance. It is no exaggeration to say that the influence of acoustics on modern physics has been very great indeed, largely through its concern with the field concept. It exhibits practically all the philosophical and historical problems discussed in this book.

Chapter One

THE METHOD OF PHYSICS

Definitions

Everyone knows that physics is a science and that it deals with things variously called natural phenomena, the material world, or properties of matter. Pressed to be more specific, the average person mentions mechanics, electricity, heat, sound, light, and atoms as concerns of physicists. But these are mere collections of names and provide little, if any, clue to what a physicist really does. The same is true of the common definition of physics as the science of matter and energy. These are concepts of physics, to be sure, but by themselves they tell nothing about the subject. It is, indeed, the province of physics to define and utilize these concepts.

A useful working definition of physics is that it is a science and that science is a method for describing, creating, and understanding human experience. But first the significance of the indefinite article *a* that precedes the word *method* must be emphasized. There are many ways of dealing with human experience, including, for example, the arts of music, poetry, painting, and drama. Science is only one way, albeit a very powerful one. Physics therefore is a rather specialized view of experience, and this must be kept in mind in the subsequent discussions of some of the arbitrariness associated with it.

The Nature of Human Experience

Experience in the sense in which we shall use it is the sum total of everything that happens to each one of us in all our waking, and perhaps even sleeping, hours, along with the reflections on these happenings made by our minds. These happenings are given various terms—sense impressions, sensations, sensory perceptions—by philosophers and psychologists. Put more specifically, we as individuals are continually seeing, hearing, touching, tasting, and smelling, and from our mental reactions to these sensations are creating ideas about the objects of our perception and their relations to each other. (This is a simplification of a very difficult philosophical problem, about which reams of philosophical literature have been written.)

Before the term experience becomes meaningful in the context in which we use it, it is necessary to postulate also that the sense impressions of different individuals agree sufficiently so that the individuals can discuss their impressions without hopeless confusion. If A could not act on the presumption that B's normal experiences in the same environment as A closely approximate A's, there could be no science and no physics. Of course, sometimes the experience of A will differ radically from that of B, because the perceptions of A or B are "abnormal." In practice, however, distinctions between normal and abnormal experience have been fairly easily established.

Not all human experience has been incorporated into physics. Physicists have been highly selective in their choice of experience for study—indeed more so than is commonly realized. Still, the domain of experience tackled by physicists is continually expanding and cannot be defined too closely. For instance, while it has been customary to take for granted that physics has nothing to do with the observed behavior of living things, scientists nowadays have no hesitation in developing biophysics and psychophysics, in which the methods of physics are applied directly to biology and psychology. Similarly, although experience connected with the transformation of one substance into another has historically been considered the province of the chemist rather than the physicist, the field of chemical physics is now flourishing, and the distinction between chemistry and physics, except for artificial administrative purposes, has become practically negligible. To attempt to delimit the domain of experience peculiar to physics here would be to place the emphasis in the wrong place. It is the method of physics that is important.

Description in Physics

The word description literally means a writing about something. It can also be used in the sense of talking about the thing being described, with the intention that the listener form some idea of it without actually seeing it himself. Description requires that the talker should, by comparison with other things alike or different, create in the mind of the listener an image of the thing described. In the science of physics, to describe experience is to talk about things observed in the talker's environment; but in physics the talking is governed by the rather stringent limitation that it must relate to the search for order in experience. The flux of experience as encountered by a young child must be a rather chaotic affair, and presumably the same was true of man in the early stages of his development. But gradually certain patterns emerged in the form of such repetitive natural phenomena as the succession of day and night, and the courses of the stars. These gave man a handle on his experience; he could build a feeling of confidence in the light of this observed order, even when it was occasionally marred by disturbances like eclipses, violent storms, volcanic eruptions, or earthquakes. The perception of experience as involving a degree of order or regularity triumphed over the more pessimistic view that one could never feel any assurance about what was going to happen next. The victory of the optimistic point of view has made the science of physics, and indeed all science, possible.

It is not only the common repetitive phenomena that seem to indicate order in human experience. Simple relations between apparently diverse experiences—between lightning and thunder, dark skies and rain or snow, the stretch of the bowstring and the speed of the arrow, the pitch of a musical note and the size of the struck object producing it, the rays of light reflected from a mirror and the size and shape of the mirror—all these relations suggest regularity in experience and stimulate the search for further examples. The curiosity thus developed has been one of the great driving forces of science and has led to the quest for relations of a more recondite kind—like those between electrical phenomena and magnetism; the size, shape, and constitution of solid objects and their ability to conduct electricity; and the velocity of light and the nature of the medium through which it is traveling.

When the properties that appear always to be related in a regular fashion can be specified with some degree of definiteness, it becomes possible to express the relation in precise language. Of course it will scarcely be worthwhile to do this unless the relation has been observed many times by many people in many different places, and general agreement has been reached about its nature. Under these circumstances the description of the relation becomes a physical law—a brief expression of a routine of human experience in the physical domain. So we have laws like Boyle's law for gases, Hooke's law for elastic substances, Ohm's law for electric currents, the law of the refraction of light, and the law of the pendulum. In the establishment and enunciation of laws of this kind, sophistication of a high order is involved. The nature of physical law will be discussed at greater length in Chapter Two; meanwhile, some obvious questions arise, and these questions can be taken up here.

The very use of the term law conjures up in the mind of the average thoughtful person the vision of a law-abiding universe that it is the obligation of the physicist to discover or, perhaps more properly, to uncover bit by bit. From this standpoint, a physical law describes a regularity in experience ordained, as it were, from the beginning, but concealed from man until clever observers and experimenters came along and detected it. It must be stressed, however, that this is by no means the physicist's interpretation of law. To him it means no more and no less than the simplest possible description of a routine of experience, amply attested by numerous experimental tests under carefully controlled conditions. That there can be no element of logical necessity in its status is made clear by the fact that no physical law as yet set up has proved to be valid over the whole experimental range of the quantities to which it relates. Boyle's law, for example, ceases to be an accurate description of the behavior of a gas at very low temperatures. Under these circumstances, of course, all gases change their state, and near the transition point between phases Boyle's law no longer functions. This difficulty led to the restriction of Boyle's law to the description of the behavior of an ideal gas, where an ideal gas is taken as a kind of theoretical extrapolation of a permanent gas (one hard to liquefy), like hydrogen at room temperature and above.

No physical law should ever be stated without an accompanying statement of its limits of application. The physicist seeks to establish laws that have the widest possible domain of applicability. At certain stages in the development of the science it was fondly hoped that more and more precise observation and more skillful selection of the variables influencing the observed behavior of physical systems would lead to laws of simple analytical form. Newton, for example, had a profound confidence in the essential simplicity of nature. Unfortunately, experience to date indicates that this confidence was not well founded, though much depends on what is meant by simplicity.

Since a law states a relation connecting ideas that refer to experience, it is clear that a law's logical status cannot profitably be analyzed before settling the question of how these ideas are actually expressed in terms of experiential data. It is necessary, in other words, to examine the nature of physical concepts and their linguistic representation in order to understand how a law describes experience. Without such an analysis words like *pressure, electrical current, force,* and *mass,* and statements about them are meaningless. The purely operational significance of concepts must be contrasted with their theoretical meaning—their relation with other concepts in the same or related domains. The view, sometimes known as operationalism, that physical ideas have no validity unless they are definitely tied to specific operations in the laboratory must be examined with some thoroughness, as must the effect of this view on the development of theoretical physics. Finally, law and theory are inextricably related to one another, for in every law there is an inevitable element of hypothesis.

The Creation of Experience in Physics

The experience that is described in physics should not be understood only as something taken in passively by the human observer, though that, indeed, is presumably how science started. People simply looked about them, tried to reckon with what they experienced as best they could, and accidentally came upon signs of order. But man eventually woke up to the fact that the experience received in this way is limited in extent and made the overwhelmingly important discovery that man himself can create experience by setting up arbitrary arrangements of objects and performing operations on them with the aim of seeing what will happen. Archimedes, for instance, carried out an experiment when, instead of simply observing that some objects float in water and others do not, he tried to find out whether there is a relation between the weight of a given object in comparison with its size and the volume of water it displaces.

Experimentation is often referred to as controlled sense perception. This conveys some of the meaning of the process but misses its most vital characteristic—the fact that it produces experience new to man. In this sense experimentation may be described as the creation of experience. Science as a method became powerful when scientists decided not to limit experience only to what is at hand but to go out and produce it. This creation of experience through the medium of experiment is going on ceaselessly in countless laboratories throughout the world and in physics has reached an advanced stage of development in terms of the complexity and cost of the equipment being used.

It is obvious even to a casual observer that experimentation is not a random affair, that physicists do not merely put things together in a haphazard way in the hope that something will emerge. The creation of experience in physics is, in fact, based on prior experience, as well as on a considerable amount of reflection on that experience. Much light is thrown on this involved process by the discussion of logical analysis of physical theories in the following chapter. But there are also psychological factors inherent in the inventiveness of man (and experiment certainly implies invention) that are not clearly understood.

To carry out an experiment a physicist abstracts from the totality of experience a certain small segment for special study. He may decide, for example, that he is interested in the nature of sound, which he believes travels through the air from one place to another by means of some sort of propagated disturbance. He has reason to believe that it takes time to travel, and he would like to determine its speed—the distance traveled in a given time. The question to be answered already suggests a program, based on prior knowledge of the measurement of speed in other situations. So a plan is devised, and appropriate equipment is chosen. The plan involves a program of operations to be carried out with the equipment. It is important to realize that this has to be in the mind of the experimenter before he can take any meaningful practical step. The final act is to perform the actual operations and to record the results in some suitable symbolic fashion. But this is not really the end after all, since the results are of no significance unless they have really created some new experience, and this can only be ascertained by examining and interpreting the results, and comparing them with the results of other persons who may have carried out experiments along the same line. Every step in the process calls for logical analysis.

Most physical experiments involve the measurement of a physical quantity. Put qualitatively and with deceptive simplicity, a measurement is an experiment or group of experiments whose aim is not merely to answer the question of how but also that of how much. It represents the attempt to satisfy the quantitative urge by attaching numbers to the results of an experiment. Number is one of the most powerful and at the same time most mysterious concepts ever created by the human mind. Its origin is presumably to be found in early man's preoccupation with the comparison of assemblages of things as well as his later desire to order things with respect to variation in physical properties—as, for example, when he sought to make more specific his assertion that one thing is larger than another.

Essentially all measurement in physics reduces to counting. But counting, except in the simple case of enumerating an assembly of entities, involves the use of a scale—a suitable physical surface on which a set of marks is inscribed in some arbitrary though definite fashion. A meter stick is a familiar illustration; others are a common thermometer, the speedometer of an automobile, or the face of a watch or clock. Attached to the marks are cardinal numbers, usually arranged in increasing order of magnitude. The spacing of the marks is arbitrary, though the most common kind of scale is the linear variety in which the number assigned to any particular mark is directly proportional to the distance of the mark from the chosen origin, usually indicated by zero. Among other important scales is the logarithmic one, illustrated by the standard slide rule, in which a large range of values of a given quantity can be compressed into a relatively small space.

To use a scale for measurement a coincidence must be established between a mark on the scale and some physical object, such as a pointer. In the measurement of the length of a table by means of a tape measure, the zero mark of the tape must be placed in coincidence with one edge of the table and the coincidence established between the other edge and a mark on the tape. In this case the other edge acts as a pointer. It literally points to the mark on the tape that gives the number of inches or centimeters in the length of the table. Most common physical measuring devices have built-in pointers. Thus that very simple but profoundly important instrument, the clock, has the hands as its pointers, whose coincidence with the marks on the face (the scale) by some arbitrarily chosen scheme, enables us to "tell" the time—to carry out the process of measuring time in any experiment. Another example of a measuring instrument is an electric ammeter, in which a needle is arranged so as to move across a scale, thereby attaching a number to the electric current flowing through the circuit in which the meter is inserted. All physical measurements involve some kind of pointer-readings, as Eddington has stressed.

A measurement determines the magnitude of a physical quantity. It is therefore essential to devise some way of identifying and talking about this quantity. This involves the notion of symbolism. A measurement could of course be described in purely operational terms, by stating that in a given case, when certain equipment was put together with a scale and pointer incorporated in it, certain readings were obtained when certain conditions were varied while others were held constant. This would describe what took place, but physicists have long since ceased to be satisfied with this way of putting the matter. To talk effectively about what has been done in the performance of an experimental measurement, it is advisable to replace the long-winded method just described by saying that the given set of operations constitutes a measurement of velocity, or mass, or pressure, or temperature. As soon as we have introduced such terms we have embarked on what is called a symbolic terminology of physics. The use of symbolism, whereby a single word or term stands for a whole series of operations, is obviously of the greatest importance in physics. It brings us face to face with the use of mathematics in a more significant way than the mere employment of numbers does. This is because symbolic terms like velocity and mass are ultimately represented by algebraic quantities that can figure in equations and be subject to the rules of mathematical manipulations.

Another problem involving measurement is commonly known as experimental error. When a physicist makes a measurement of a particular quantity, he usually repeats his observation several times in order to make sure he has made no mistake. Now it almost invariably happens in this process that he will not obtain the same number each time. The question then arises as to which number he shall take as representing the result of the measurement. Unless he can reach a decision on this point, the meaning of the measurement is put in jeopardy, or at the very least would seem to lead to more complications than it is worth. Hence a procedure must be adopted to assign a final value to the measurement. The theory of errors by which this is done is a somewhat arbitrary, but on the whole plausible, technique. The word arbitrary should be emphasized, for it symbolizes a key notion in the whole process of building a science like physics and is a notion we shall encounter again and again as we proceed.

Some humanists resent the general idea of the creation of experience in physics. They feel that it distorts the natural world it is the duty of the scientist to study and report on. But the creation of experience is what constitutes the coming to grips with the world as the scientist conceives of it, and it must be remembered that humanists also create experience after their fashion. Nevertheless, criticism of the creation of experience can serve a useful purpose, if it encourages physicists to examine from the pragmatic standpoint what limits, if any, should be placed on the domain of physical experience. There does not appear, however, to be any aspect of human experience that is logically completely immune from examination by the physicist.

Understanding in Physics

Physicists describe experience, they create experience by performing experiments, and having accomplished these essentials, they seek further to enlarge experience by understanding what they have done. To a physicist, understanding experience involves the development of physical theory. A theory is an imaginative construction of the mind that employs ideas suggested by experience and also by arbitrary notions whose origin it is difficult to trace; together these ideas and notions form a kind of mental picture of things as they might be. If a theory is to be of any use, it must be possible in reasonably unambiguous fashion to draw consequences from it that can be identified with actual experience. These consequences must presumably be expressible in terms of laws that are equivalent to the descriptive laws discussed earlier in this chapter. Moreover, if it is really to be successful, the theory must predict results that have not hitherto been experienced; the suggested experimentation when performed must then substantiate the prediction. Of course, even if the results are not wholly in agreement with the prediction, this does not necessarily mean that the theory has been wholly useless, as we shall see in our subsequent investigation.

As a simple illustration, consider the nature of light, a segment of experience of enormous significance to man and one that has been the subject of observation, experimentation, and speculation for centuries. Regularities that have long been known to exist have led to the laws of reflection and refraction; the laws of diffraction by obstacles, slits, and gratings; and the laws of polarization. The problem is to understand these manifestations of order in this domain of experience. To this end, numerous theories have been devised, of which one generally successful one is the wave theory. This assumes that light is propagated as a wave in a medium, a wave being simply a disturbance of any kind that will not stay put at its point of origin but persists in traveling through its medium. For example, when a stone drops to the surface of an otherwise smooth body of water, the surface is locally disturbed, and a ripple or succession of ripples travels out in all directions from the place where the stone entered the water, carrying the original localized disturbance to other parts of the surface. This surface water wave thus results from the transmission of motion over the surface of the water, without the particles of water themselves moving across the surface; it is the shape of the surface that moves. The wave theory of light similarly consists of the assumption (among others) that a luminous object or source of light produces some kind of disturbance in a medium whose nature and properties are not immediately specified (a difficult feat, since it is known that light travels through a vacuum in which no matter as commonly understood resides) and that when this disturbance reaches the eye it produces the sensation of vision. It is perfectly possible to deduce from the fundamental wave hypothesis that waves in general can be reflected from solid surfaces on which they impinge. They can also be refracted—the direction of propagation of the wave can be changed on transmission across the boundary that separates media of different properties, such as different wave velocities. It is possible to predict further that the direction of propagation of a wave will be altered in passing around an obstacle or through a hole in an otherwise rigid wall, that is, it will be diffracted. Light shows all these properties, and the precise laws governing them are found to agree with those deduced logically from the fundamental assumptions of the wave theory of light. To this extent we may say that the wave theory of light provides an adequate understanding of observed human experience in the domain referred to as light.

But this, as has been suggested above, is hardly enough. Confidence that the wave theory is really a good theory came only when predictions based on it led to experience about light not hitherto within human ken. One example is that of the French physicist Augustin Fresnel, who deduced from the wave theory that a beam of light should bend around a circular obstacle and should do so in such a way that though most of the far side of the obstacle would cast the usually expected shadow, the constructive interference of the light waves on the perpendicular to the obstacle through its center should lead to an illuminated spot in the middle of the shadow. At the time when this was predicted, no one, so far as is known, had ever observed such a thing, and indeed it is still not an easy phenomenon to observe except with special optical equipment. The prediction at once suggested appropriate experimentation that when carried out, completely substantiated the deduction from the wave theory. A more elaborate prediction came from the application of the theory to the transmission of light through transparent crystals of the kind called double refracting. Sir William Rowan Hamilton, the celebrated Irish mathematician, astronomer, and physicist, predicted that in certain cases the refracted light in such a crystal would form a cone of rays. He called it internal conical refraction. Once again the effect had never been observed and seemed to some to be rather fantastic. But it was ultimately shown to exist experimentally.

The history of physics is full of such examples of successful prediction of new experience by theories. One thinks, for instance, of the prediction of the existence of the positive electron or positron by Paul Adrien Maurice

Dirac's version of relativistic quantum mechanics or of the existence of the red shift in spectral lines in an intense gravitational field like that of the sun, as predicted by Einstein's general theory of relativity. Still more famous and indeed awe-inspiring in its ultimate impact on technology is Einstein's prediction of the relation between mass and energy ($E = mc^2$), based on the special theory of relativity. Other illustrations will be discussed in subsequent chapters. The point here is that great intellectual power is inherent in the predictive capacity of a successful theory and that such theories play a vital role in the creation of new experience.

An emphasis on predictive ability as the criterion for success oversimplifies the appraisal of physical theories. If successful prediction were the sole criterion, physical theorizing might well be reduced to a gambling game in which a lucky guess wins the prize. Moreover, it may be asked, what is to be done if two quite different theories predict the same result, and the prediction is verified? Could the physicist be satisfied with this situation? The answer is that theories must be appraised on other bases as well. Among these bases are judgments on: the choice of concepts or constructs acceptable in terms of known experience; the limitation in the number of independent constructs employed; the mathematical elegance and rigor in the formulation of the theory; and the comprehensiveness of the theory—the extent of the domain of experience in which it works. All these judgments play a role in any valid appraisal of the contribution of a theory to the understanding of experience.

The problem of the nature of a physical theory is therefore not so simple as an undiscriminating interpretation of this section might lead one to believe, and still further questions arise. One example is the relation between law and theory. In this preliminary discussion we have treated physical law both as a brief expression, almost always in analytical form, of a routine of experience and as a logical deduction from the hypotheses of a theory. We must ultimately justify the choice of the same word in these two meanings, the first descriptive and operational, the second theoretical. We must face the question of whether a statement that is a deduced law in one theory can serve as a postulate in another, and in the chapter that follows these and related questions about physical theorizing will be examined.

Chapter Two

THE LOGICAL STRUCTURE OF PHYSICAL THEORY

Analysis

Having discussed the nature of physics as a science and emphasized the significance of physical law, experiment, and theory, we now find it necessary to examine in greater detail the nature of a theory, for it is the key to understanding physics and indeed all science.

There are two fundamental approaches we can take, the psychological and the logical. The first would investigate how clever people come to create theories. This may properly be called a part of the psychology of physics. It is a very difficult subject, and it is therefore not surprising that its exploration has led to little of positive value. One might suppose that creative physicists would have been at pains to indicate at some length how they arrived at the basis of their theories, but they have not often done so or at any rate not in such fashion as to provide definite recipes for the construction of theories. Jacques Hadamard has presented interesting illustrations of mathematical discoveries by Jules Henri Poincaré and others, but it must be confessed that from any practical standpoint the results are disappointing.[1] The same is true of the more recent ambitious attempts by Abraham Moles and René Leclercq.[2] There are hints, but a much more thoroughgoing psychological investigation is needed.

Here we shall take the second approach and make a logical analysis of physical theory and try to grasp the significance of its various components and their relation to each other. While this will of course not provide a recipe for the concoction of theories, it will set forth the structure of a theory in such a fashion that we can more readily understand the various questions of a methodological nature that may appropriately be raised about physical theories in every field. It should give us a pretty good notion of what physicists mean when they say that theory provides an understanding of a part of human experience.

For convenience a physical theory can be logically analyzed in terms of the following schema:

1. Primitive, intuitive notions
2. More precisely defined constructs

3. Postulates or hypotheses connecting the constructs
4. Deduced laws
5. Experimental testing of the deductions

Let us try to make this schema clear. The best we can do in the way of grasping the meaning of anything is to talk about it in words that compare it with something supposedly more familiar. To revert for a moment to the illustration of the wave theory of light in the previous chapter, we seek to understand light in terms of waves, and the picture that the mind conjures up is one of waves on the surface of water—waves as a visual experience. To get a clear idea of even a surface water wave, however, implies our ability to recognize motion as the displacement of something through space as time goes on. Now space and time are primitive and rather ill-defined notions that are nevertheless fundamental for physical theorizing. One cannot do business without them. But the creative physicist generally takes them for granted. He thinks that every intelligent person will be willing to admit a familiarity with the concepts of space and time, and so in building theories the creative physicist does not hesitate to introduce such primitive notions as purely intuitive and not subject to further question. These constitute the first class in the preceding logical schema. Some other illustrations are the concept of a material particle, the concept of a field as a region in which each point is characterized by a definite value of some physical quantity (temperature, for instance), the idea of causality, the concept of probability that is so useful in statistical theories, and the notion of symmetry.

It is obvious that though these ideas are usually taken for granted by the framer of a physical theory, they necessarily pose many serious problems for the philosopher of science, who looks with suspicion on too many undefined terms in any logical structure. At the same time even the philosopher has to admit that definition of anything ultimately means talking about it in terms of other things and that sooner or later one runs into undefined terms. Any dictionary illustrates this difficulty: all definitions reduce to circular definitions, as must be the case in any finite domain of human discourse. Euclid effectively recognized this difficulty with such undefined terms in his geometry as point, line, and plane. The definitions of these concepts, such as "a point is that which has no parts," are really no more than reminders that the concept ought to be pictured intuitively. In no branch of science can we ever get away from this fundamental difficulty: it faces the human race wherever and whenever the attempt is made to talk in terms of ideas instead of mere objects that can be pointed to.

We escape from this somewhat uncomfortable part of the logical structure of physical theory by passing on to the problem of the building of physical constructs.

Constructs and Symbolism

Using the primitive, intuitive concepts just discussed, the physicist proceeds to manufacture more precise concepts or "constructs," to use a term popularized by Henry Margenau that emphasizes more clearly what is meant.[3] A simple example is the vector displacement of a particle. The position of a particle in three-dimensional space is represented by the vector or directed straight line from the origin of the co-ordinate system to the position of the particle. Displacement is defined as the vector difference between two successive position vectors. This difference between two vectors is further defined as the vector that when added to one of the vectors, produces the other as a resultant. This in turn demands a definition of what is meant by the sum of two vectors.

Even the assignment of meaning to a very simple construct like displacement requires a considerable amount of elaborate discussion and assumption, and that it involves a certain degree of arbitrariness is clearly evident in the preceding example. It is by no means certain that we can legitimately expect that the vector difference of two position vectors will serve as an adequate representation of the displacement of the corresponding particle. In fact it does not require much knowledge of physics to see that as the two positions of the particle get farther apart the difference between the corresponding position vectors becomes a less likely representation of the displacement of the particle. Hence arises the construct of differential displacement, in which the length of the difference between the position vectors is a small fraction of that of either vector. Already the need for something like the differential calculus creeps into the picture.

Once over the initial hurdle, things seem to go more smoothly in the construction of a concept like velocity (the rate of change of displacement with time) and acceleration (the rate of change of velocity with time). This smoothness is of course only a kind of illusion and is anything but apparent to the uninitiated. There are at least two aspects of the question that demand more careful examination. These are the use of symbolism in the formation of constructs and the distinction between the operational and theoretical aspects of definition.

A symbol is a sign used to represent an object, an idea, or a situation. Thus a name symbolically designates a person or thing. But it can also represent an idea like love, or justice, or mass, or energy. All the terms in a dictionary are essentially symbols in the language in question. For reasons of economy, it is often customary, particularly in a science like physics, to replace actual linguistic names by single letters, as m for mass, V for volume, or p for pressure. The further and even more significant advantage of this is that if the thing being symbolically represented has a quantitative aspect and is therefore measurable, the symbol can take on numerical values and enter into algebraic relations with similar symbols. One can then apply all the paraphernalia of mathematics to it. This is a very powerful and revealing procedure. It enables us to express the content of physical laws in mathematical form, which is not just a matter of economy but also allows the use of mathematical manipulation in the deduction of new consequences from any statements in which the symbols enter.

A particularly important point emerges when the construct we are trying to define is a quantity to which a number may be attached, like mass, volume, or pressure. An electron, on the other hand, is not a quantity, and yet it is a perfectly good example of a construct. No number may be attached to it in itself, though its various properties—size, mass, charge, and velocity—are all quantitative under certain circumstances. In fact, it is hard to think of a physical construct that is not somehow concerned with numerical significance.

The assignment of a number to a construct or to a property of a construct involves what is called a measurement. This is a series of operations like those associated with any physical experiment, but with the added requirement that the equipment contain a scale and a pointer. The process of measurement is one with which nearly everyone is familiar, since all persons in countries of any pretense at all to civilization use it daily. In most cases the pointer is an actual solid moving indicator like a needle; but it does not have to be. Thus in an ordinary mercury-in-glass thermometer the pointer is the surface of the mercury. Its coincidence with a mark on the scale is a measure of the temperature of the environment in which the thermometer is placed. In a wall galvanometer a spot of light can be the pointer. The varieties of meters for the measurement of quantities like voltage, current, resistance, stress, pressure, speed, or intensity (of light or sound) are legion, but they all involve a pointer and a scale. The latter in general will have numbers assigned to the marks, but this is not universally the case. For instance, in the conventional beam-and-pan balance used for the measurement of mass, the scale is a null one in which the pointer must merely come to and stay in coincidence with a single mark in order to carry out the measurement. The number in this case must come from the numerical values of the weights put in the pan in order to achieve the balance.

The construction of a scale is a more or less arbitrary affair. The spacing of the marks need not be uniform, though most commonly used scales are linear. The precise spacing is largely a matter of convenience, though it is necessarily associated with the type of measurement being performed. Some thinking about the whole phenomenon must therefore precede the experimental measurement before the latter can be expected to be meaningful—in other words, a good deal of theory enters into every measurement.

The definition of a physical construct that represents a quantity must provide directions as to how the quantity is to be measured. This is the operational, or epistemic, aspect of the definition. A definition that fails to have this side to it evidently can be of little use in categorizing a measurable quantity. To define mass, for example, as the amount of matter in a body is meaningless, since this definition does not convey any recipe for assigning a number to mass. On the other hand, a purely epistemic definition is not always wholly adequate either. A good operational definition of mass can be constructed out of the ordinary directions for the use of a beam balance, but this hardly conveys the meaning of mass in the theory of mechanics, in particular its relation to the other important constructs of theory (displacement, velocity, acceleration, force, and energy). To obtain a complete definition of mass we must introduce some theoretical aspects. The word constitutive has been applied by Margenau and Northrop to these.[4] In simple terms this aspect of the definition of mass figures in the postulational equation (force equals the product of mass and acceleration).

To summarize, every physical construct that represents a physical quantity subject to measurement must contain both epistemic and constitutive aspects in its definition. It is true that certain physical theories contain constructs representing quantities not subject to direct measurement. Such a construct might be the radius of an electron orbit in the simple Bohr model of an atom. Since there is no direct laboratory experiment by which this quantity can be measured, its definition will be wholly in terms of other quantities in the Bohr theory and hence wholly constitutive in character. In general, the more elaborate and sophisticated a theory is, the more purely constitutively defined constructs it will contain. Of course, these quantities must disappear in the laboratory equations resulting from the theory. Not surprisingly, the most numerous and striking examples of purely constitutively defined constructs are to be found in modern physical theories like the quantum theory in its various phases and applications.

Postulates

The next step in the logical schema of a physical theory is the nature and place in the picture of the fundamental postulates. We have already looked in Chapter One at the general problem of the invention of clever ideas in the form of pictures of experience as it might be—pictures that, if accurate, are representations of actual experience in the observable world. But it is important to re-emphasize the key position of hypotheses in physical theory. The very manufacture of constructs as discussed in the preceding section involves hypothesis: one cannot build the construct of mass, for example, without making a number of assumptions about the motion of particles in each other's vicinity. But from the standpoint of the logical structure of a theory it is customary to place special stress on the assumed relations among the constructs from which deductions can be made. Thus in the theory of mechanics in the version commonly associated with the name of Newton the equation of motion ($F = ma$) is the key hypothesis. This has been much misunderstood in the teaching of physics. Many have treated it as an experimental law like Boyle's. But the equation of motion clearly does not fall into this category, since analysis shows that force, mass, and acceleration cannot be defined independently as measured quantities in such a way that the equation describes all the cases of motion. The equation of motion is not a laboratory equation as are the laws of Boyle, Ohm, Joule, or Snell. However, if we assume that it has universal validity, we can by appropriate choice of the force function F derive equations that directly describe types of motion encountered in our experience. One such equation is the famous law of freely falling bodies, in which the distance fallen by a body dropped from rest at a point not too far from the surface of the earth is equal to a constant times the square of the time of fall. The constant in this case is one-half the acceleration of gravity (usually represented by the symbol g), with the numerical value approximately equal to 980 centimeters per second per second.

Similarly, Boyle's law may be deduced by a mathematical analysis based on the postulates of the molecular theory of gases. The basic postulate is the existence of the molecule as a small particle whose motion obeys the principles of mechanics. But one must be more definite than that in the further assumption that a gas is constituted of a very large number of

molecules moving in all directions with all sorts of speeds and colliding
with the containing walls. Still another hypothesis is that the average rate
of change of momentum of the molecules colliding with unit area of the
walls per second will be a good representation of the pressure exerted by
the gas against the surface of the vessel that contains it. It then remains to
calculate this average rate of change in momentum as a function of the
average of the molecular velocity squared. This may be done in relatively simple, approximate fashion or it may be carried out with much
more elaboration to take care of such complications as the different directions from which the molecules strike the walls. However, in all cases the
result is the same: the average change in momentum per second per unit
area is inversely proportional to the volume of the containing vessel. The
coefficient of proportionality is two-thirds of the total average kinetic energy of the whole collection of molecules confined in the vessel. In order
that Boyle's law shall result, it is necessary to make the further assumption
that the temperature of the gas is somehow connected with the mean kinetic energy of the aggregate of molecules, so that when the latter is constant, the temperature is constant. With this additional assumption, Boyle's
law follows from the relation whose derivation has just been traced. If we
are willing to specify the relationship between temperature and mean kinetic energy more precisely, we can derive a more general law. It is generally assumed that the mean kinetic energy per molecule is directly proportional to the absolute temperature of the gas, and that the coefficient is
equal to three halves of a universal constant of nature, called the Boltzmann constant and usually represented by the letter k. With this postulate
the relation we have deduced becomes the more general one

$$p V = NkT,$$ (Equation 1)

where p represents the pressure, V the volume, T the absolute temperature,
N the total number of molecules, and k is the Boltzmann constant (1.38×10^{-16} erg/°C per molecule).

It may strike the thoughtful observer as a bit disconcerting that so much
has to be assumed in order to lead to the general equation of state of an
ideal gas (Equation 1). This feeling would be shared by the physicist if all
that could be done with the postulates were to deduce Equation 1. In this
case he would not feel that he had a valid physical theory, for the agreement with experience might be merely accidental. Since, as Chapter One
suggests, a theory is of little value unless it leads to many consequences
that are then shown to be in agreement with observation, the funda-

mental postulates of the molecular theory of gases should lead not only to the equation of state but should also predict without too much more elaborate machinery that an ideal gas possesses viscosity (the tendency of two parallel moving layers of gas to decrease each other's relative motion). The same theory should be able to predict the dependence of the viscosity on the temperature and pressure. Further, it should make possible the evaluation of the numerical values of the specific heats of the gas at constant pressure and constant volume. And so one might go on with other properties of an ideal gas that the molecular theory should predict. As a matter of fact, it does pass these various tests pretty well. This single set of fundamental postulates manages to imply a host of consequences not immediately apparent from a casual inspection but that can be developed with the use of a little imagination plus the use of appropriate mathematical analysis. This is the challenge of a plausible theory: to squeeze out of it as much as one can.

This squeezing process can have two consequences. Either every deduction from a theory will be found to agree with experiment or somewhere along the line a result will be obtained which disagrees with observation. It must be admitted that in practice the latter almost universally occurs. Numerous examples are to be found in all physical theories. Thus, in the molecular theory of gases one can readily deduce a simple relation between the viscosity η, the specific heat at constant volume C_v, and the thermal conductivity θ:

$$\theta = \eta C_v. \qquad \text{(Equation 2)}$$

While experience agrees with this as far as order of magnitude is concerned, the ratio of $\theta/C_v\eta$ is usually closer to 5/2 than to unity for simple monatomic gases.[5] Often the degree of disagreement between a deduction from a theory and actual observation is far greater than this—as in the classical theory of mechanics, which, in its simplest form, implies that a pendulum bob once set swinging and let go will oscillate forever, whereas we all know that the contrary is the case. When such a situation arises—and it has, as a matter of historical fact, arisen in the exploitation of every physical theory that has ever been devised—the physicist can react in either of two ways. He can conclude that the basic postulates of the theory are not plausible, that the theory is all wrong and should be discarded completely. On the other hand, if the number of experimentally confirmed laws deduced from the theory is considerable, he will probably reject this attitude as unduly pessimistic. He will feel that the foundations of the theory are still sound and that the agreement with experiment is not fortuitous but significant. The usual course of action in this case is to attempt to make small changes in the assumptions to see whether the results of the unsatisfactory application cannot be modified in the right direction without leading to incorrect results in the deductions, which originally agreed with experience. A very large part of the development of theoretical physics is devoted to this sort of activity. It may take genius to invent the principal idea or set of ideas behind a given theory, but it still demands much ingenuity and hard work on the part of the less distinguished workers to tidy up the rough spots.

Of course, cases occur in which no amount of tidying up seems to accomplish the desired end. Much effort was expended in the late nineteenth century in trying to deduce from the classical theory of electromagnetic radiation (due essentially to James Clerk Maxwell) the law that describes the distribution with respect to frequency of the energy of the radiation emitted from a hot black body. According to this experimental law the radiation energy rises from zero in the limit of very low frequency to a maximum at a frequency that depends on the temperature and then falls asymptotically to zero for very high frequency. Classical thermodynamics as applied by the German Wilhelm Wien yielded a formula that fits the experimental curve rather well for very high frequencies.[6] It even predicts the existence of a maximum, but its behavior at very low frequencies is in complete disagreement with the experimental results.

From the field of statistical theory, a somewhat different approach, made by Lord Rayleigh and Sir James Jeans, produced an equation in good agreement with observation at very low frequencies, where the Wien law breaks down, but in hopeless disagreement at high frequencies.[7] Moreover, the Rayleigh-Jeans law does not predict a maximum in the curve at all. In order to solve this puzzle Max Planck discarded some very fundamental assumptions of the classical theory and introduced an entirely new concept, that of the quantum of energy.[8] A new theory was then developed, in 1900, with incalculable consequences for the development of physics and its associated technology. The question arises of whether the old classical theory was completely discarded. Strictly, the answer is that it was not. Essentially, what has happened is that the classical theory has come to be considered as a kind of limiting case of quantum theory (it won't work the other way around) applicable to certain physical situations in which atomic constitution effects may be neglected. The case of the radiation of radio waves from an antenna is an example; it can, of course, be treated by the quantum theory, but since here the new theory merges with the old, it is unnecessary.

These illustrations indicate that though the logical scheme on which we are hanging our exploration of the nature of physical theory is a convenient peg, it conceals many somewhat tricky points. All is not so logically straightforward as the original discussion might seem to imply. Questions are bound to arise. For example, what if two different sets of postulates predict the same experimental results? This has happened on several occasions in the history of physics. Until 1850 the prevailing theory of heat was the caloric theory, whose basic assumptions were that heat is an indestructible substance and that materials become warm by absorbing caloric and cool by releasing it. It was realized that caloric was different from ordinary matter in that it possessed no weight that could be detected. The theory was successfully applied to account for the known thermal properties of matter, such as specific heat and latent heat.[9] Nicolas Léonard Sadi Carnot used the theory in his attempt to understand the motive power of heat as manifested in heat engines, thus foreshadowing, around 1820, the second law of thermodynamics.

But even before the eighteenth century had come to a close, grave doubts about the caloric theory had already arisen. In 1795 Benjamin Thompson, better known as Count Rumford, had performed his famous experiments on the heat produced in the process of boring metal for making gun barrels and had reached the conclusion that the caloric theory was inadequate to account for the results.[10] From the fact that merely by the expenditure of mechanical work one can seemingly produce an arbitrarily great quantity of heat (enough to boil an arbitrarily large amount of water), he concluded that it was more reasonable to associate the production of heat, as something evident to the senses, with the doing of mechanical work than with any hypothetical substance that could, so to speak, be squeezed out of the metal by friction. This alternative view of heat had indeed been put forward by earlier investigators. To us today it not only appears more reasonable than the view of the adherents to the concept of caloric, but we accept without question the mechanical theory of heat that Rumford supported. In his day the calorists had answers to his objections that seemed sufficiently plausible to keep the old theory afloat for another half century.

What it comes down to is this: No experiment has ever really demonstrated the caloric theory to be false in the sense that the theory could not accommodate itself to the result—albeit with some emendations that tend to overload it with *ad hoc* assumptions. On the other hand, the mechanical theory of heat associates heat with the mechanical energy of the particles that, according to the atomic hypothesis, go to make up every portion of matter. It has the great merit of extending the concept of energy (originally restricted to problems involving the motion of large-scale objects) to problems in which no obvious motion is observed. The generalization of the concept of energy to encompass thermal phenomena took a relatively long time, largely because of the tenacity with which reputable physicists clung to the caloric theory with its essential postulate that heat substance is indestructible and can merely pass from one material medium to another. Today physicists take it for granted that the generalized concept of energy is the most important construct in the whole of physical science and probably of all science.

The replacement of one theory by another is not a clear-cut decisive affair that takes place overnight because one fails entirely and another succeeds in meeting all experimental tests. It is a highly complicated process in which many factors enter, among them taste, judgment as to whether the old theory can be patched up to meet the demands of newly acquired experience, and grasp of the possible relations between the theory and theories of related physical phenomena that make it possible to proceed in the direction of greater unity and generality. Actually the process of theory building and modification has many elements that relate it closely to artistic creativity.[11]

The next step in our discussion requires a closer look at the nature of the postulates at the basis of physical theories. Where do they come from? A theory begins as an act of imagination; it is in essence the creation of a picture in the mind of the scientist. This introduces the problem, but hardly solves it, for how does the physicist get the ideas that form the basic elements of the picture? As we have already suggested, this is a problem in psychology and one to which too little attention has so far been devoted. Too few scientists have ever divulged in any detail the processes they have gone through in setting up their theories, and even when they have discussed the matter, what they have written has not usually been helpful. This may be due partly to a feeling that the line of thought being followed was so straightforward and compelling that it should appear obvious without further comment to all thinking persons. It may also be due to a feeling that inspiration is not discussable. It is true that some very famous scientists have written about flashes of illumination that came to them when, after long periods of unsuccessful conscious thought, they had attempted to forget the matter at issue in the hope that the subconscious would take over. Such stories are suggestive but do not really instruct. Perhaps there is no instruction to be had in such matters, except by close association with the creative scientist. This is, however, possible for few. Psychologists acquainted with physics should give the problem greater attention.

One thing is sure. The thoughts that develop into valuable physical constructs come to those who are equipped to receive them by thorough preliminary preparation in the subject matter in question. It is obviously impossible to start to theorize on a branch of physics about which one knows nothing, and it is equally clear that every theorist inevitably is influenced in the creation of new ideas by the ideas he has obtained from others in his previous study.

This suggests two considerations. The first, manifested in the early stages of the development of what we may call modern physics (physics from the time of Galileo), is the tendency in the construction of physical theories to use ideas very closely suggested by, and associated with, the phenomena to be described and to be theoretically understood. The second, evident throughout the history of physics, is the tendency toward the use of analogy (the explanation of some one phenomenon in terms of ideas first introduced to account for another). A good example of the first tendency is presented by Galileo in his discussion of naturally accelerated motion in his *Dialogues concerning Two New Sciences:*

And first of all it seems desirable to find and explain a definition best fitting natural phenomena. For anyone may invent an arbitrary type of motion and discuss its properties; thus, for instance, some have imagined helices and conchoids as described by certain motions which are not met with in nature, and have very commendably established the properties which these curves possess in virtue of their definitions; but we have decided to consider the phenomena of bodies falling with an acceleration such as actually occurs in nature and to make this definition of accelerated motion exhibit the essential features of observed accelerated motions. And this, at last, after repeated efforts we trust we have succeeded in doing. In this belief we are confirmed mainly by the consideration that experimental results are seen to agree with and exactly correspond with those properties which have been, one after another, demonstrated by us. Finally, in the investigation of naturally accelerated motion we were led, by hand as it were, in following the habit and custom of nature herself, in all her various other processes, to employ only those means which are most common, simple and easy.[12]

Galileo clearly expresses the conviction that in dealing with any natural phenomenon on a theoretical basis the proper procedure is to utilize constructs directly suggested by the phenomenon itself. He seems to have been convinced that the study of motion would be more successful if he were to confine his postulates to assumptions about observable constructs like velocity and acceleration rather than to introduce vague ideas like impetus in connection with the causes of motion attributed to pushes and pulls and so on. Evidently, Galileo thought he could get something of significance out of the simple postulate that the free vertical fall of a particle near the surface of the earth takes place with constant acceleration. As everyone knows, he succeeded, for this hypothesis leads to the actually observed law of free fall (distance fallen from rest proportional to the square of the time of fall). He must further have asked himself whether any other illustrations of motion exist in which the acceleration is constant. This in turn may have led to the assumption that if an object is propelled in a direction making an arbitrary angle with the horizontal, its motion can be divided (in the mind) into two parts: one along the horizontal with zero acceleration (constant velocity) and the other along the vertical with nonvanishing but constant acceleration. Working out the consequences of this assumption led to the well-known parabolic path for a projectile in a resistanceless medium.

The ingenuity of Galileo's approach has long been recognized as a mark of his genius. To us today it seems to be a rather obvious approach, but it did not seem so to Galileo's contemporaries. It was based on the concept of instantaneous velocity, and without this concept it would have been meaningless. Galileo convinced himself of the utility of believing that when an object falls from rest its velocity of fall passes continuously through all values from zero to the one it has at any particular instant subsequent to its release. To him this seemed a natural idea, and he took great pains to make it clear with an elaborate explanation and defense. But it was not clear to his contemporaries, and they probably felt that in introducing it he was departing from his own criterion of sticking closely to observed phenomena in his reasoning about motion.

Like Galileo, the creator of a powerful and successful physical theory may think he is sticking to the observed facts when he is actually introducing a rather idealized construct. This was certainly true of Galileo's invention of the concept of instantaneous velocity. When one uses such an idea long enough, one finally gets into the frame of mind in which it seems second nature and hence directly suggested by experience. Numerous other examples of the tendency to introduce constructs that seem to the physicist to suggest themselves naturally from the phenomena to be described and understood will occur to the thoughtful reader. Among them is density, or the mass per unit volume (specific mass), as a means of distinguishing physically between different materials that may look alike. Other examples are the frequency of the vibration of a sound-emitting body as a physical representation of the heard pitch and the index of refraction of a transparent substance as a means of specifying its behavior with respect to the transmission of light. As in the case of the idea of instantaneous velocity in Galileo's mechanics, we must be careful not to exaggerate the significance of the word natural in connection with such constructs. They will not seem natural to everyone, and much imagination usually goes into their concocting. Nevertheless, they have a pictorial, or *anschaulich,* character that is not shared by the concepts, say, of modern quantum mechanics; its state function and operator constructs offhand seem to have no connection at all with the observed phenomena, like the frequency or intensity of spectral lines, that they are intended to explain.

Inventors of physical theories from the earliest times to the present have had a fondness for using analogy, sometimes to the disadvantage of a really successful description of experience. By "analogy" we mean here the description of a given phenomenon in terms of ideas that have proved successful in the description of another phenomenon that, as far as all observation goes, is different. Thus consider the concept of physical substance as a symbolic means of describing the obvious properties of observed matter through its density, hardness, color, and ability to change its state (as from solid to liquid) without loss in total amount. Such a concept certainly seems to be very useful to any physical theory of the properties of matter.

It is small wonder that, in constructing a theory for a mysterious phenomenon like heat, physicists should have been tempted to try to interpret heat as merely the manifestation of another variety of substance, a fluid; for like a fluid, heat possesses the ability to flow easily from one point to another. The material, or substantial, theory of heat, the basis of the caloric theory of heat, therefore arose as an analogy to the substantial theory of matter. Of course, it was realized that heat as substance must be different in some respects from material substance. Heat has no weight, for example. But the believers in the use of analogy were not dismayed, for they were dealing in analogy but not identity; things could be similar but not identical, and the similarity could contribute to an understanding of the new phenomenon. Analogy is an illustration of the urge to describe all phenomena with as much economy as possible, with the invention of the minimum number of new ideas. This is a laudable aim and has assisted to a certain extent in the advancement of physics. But from another point of view it has hindered progress. By trying to force new experience into the mold of old theory, the physicist has often had to adopt some rather far-fetched devices. This was certainly the case with the caloric theory.

The twentieth century has seen a change in viewpoint both with respect to the use of analogy in the invention of physical theories and in the sophistication of the constructs employed. The modern physicist has come to believe that whereas analogy is a useful device, discretion dictates that it be limited to cases in which the mathematical form of the description can be shown to be similar. To illustrate, it is well known that the differential equation of motion is linear and of the second order if a mass particle is displaced from its position of equilibrium under the action of a force that seeks to restore equilibrium and varies directly as the distance separating the particle from the position of equilibrium. The second-order term is the mass of the particle multiplied by the second derivative of the displacement with respect to the time, while the zeroth-order term is the product of the stiffness and the displacement. Since dissipation always enters, it is most simply represented by a term directly proportional to the velocity of the particle (the first derivative with respect to the time) with a coefficient equal to the damping constant. The solution of this equation (with the right-hand side placed equal to zero, corresponding to the particle being exposed merely to the restoring and damping forces) yields—for certain relations between the mass, stiffness, and damping coefficient—the well-known damped sinusoidal oscillation of the particle about its position of equilibrium. But now take an electric condenser that is connected in series with an electrical inductance (a coil of wire, with or without a magnet core) and an electrical resistance. Place this condenser across a source of electromotive force, give it a charge, and then separate it from the source and allow it to discharge through the inductance and resistance. The problem is to determine how the charge then varies with the time. Application of the well-known laws of electricity leads to a second-order linear differential equation for the charge as a function of time. When this is solved, it turns out that for certain relations between inductance, capacitance (of the condenser), and resistance, the charge oscillates sinusoidally—there is an oscillatory discharge of electricity through the circuit.

In comparing the mechanical and the electrical oscillations, no one can pretend that the mechanical oscillator looks like the electrical oscillator. A coil of wire certainly does not resemble physically a block of metal like a pendulum bob, nor does a condenser remind one by its appearance of a spring. Similarly an electrical resistance, which again is just a piece of wire, does not look like a vessel full of glycerine that resists the motion of the mass of the mechanical oscillator. The two oscillators are analogous simply because the behavior in each case is described by a differential equation of precisely the same type. The analogy has a purely mathematical base and nothing more. However, in the light of this, physicists have considered it valuable to think of the mass of the mechanical oscillator as analogous to the inductance of the electrical oscillating circuit, the stiffness of the mechanical spring as analogous to the reciprocal of the capacitance of the electrical circuit, and the mechanical damping coefficient as analogous to the electrical resistance of the circuit. Once this analogy has been set up, it becomes possible to introduce a visualization of either the mechanical oscillator or the electrical oscillator in terms of the other.

Since the great development of electric circuit theory has made it possible to design, construct, and utilize electrical circuits in a bewildering array of arrangements, and to be assured of their behavior in practice, it has been tempting to draw conclusions about the behavior of analogous mechanical oscillating devices purely from the electrical results. For example, if the desired combinations of mechanical mass, stiffness, and resistance elements are set down on paper and each is replaced by its equivalent electrical analogue, the appropriate conclusion about the behavior of the mechanical "circuit" can be drawn from the properties of the electrical circuit. Now it might be objected that since the analogy is wholly mathematical, it is really foolish to expect any value from it at all; presumably the mathematical details for the mechanical "circuit" can be worked out directly just as readily as for the electrical circuit. However, this overlooks the fact that clever people have conducted experiments with electrical circuits for a long time. These circuits are relatively easy to put together and to study experimentally. One therefore gets a "feeling" for them. And so, when an electrical expert has occasion to work out the behavior of a combination of mechanical components, he naturally replaces it with the equivalent electrical network about which he can think more efficiently. In this way analogy, even when it is based entirely on mathematical considerations, pays off in practice.

The use of analogy in physics is closely connected with the employment of pictorial models. An example is the Rutherford atom model, in which the atom is likened to a miniature solar system, the sun being replaced by a massive, positively charged nucleus and the planets by a series of electrons that revolve about the nucleus in orbits that are perturbed ellipses. With this model many properties of atoms corresponding to the various elements of the periodic system can be predicted. It must be confessed that such models usually break down when pressed too far. This certainly occurred in the case of the Rutherford planetary atom model, which needed the introduction of some new and quite unclassical assumptions (those of Bohr) to make it work. The elastic-solid model for the propagation of light is another illustration of the same sort of difficulty. In this case the model gave so much trouble in its details that in spite of its attractive pictorial features, it was finally abandoned in favor of the electromagnetic theory. The Bohr-Rutherford atom model has also been displaced by quantum mechanics, and this seems to be the ultimate fate of all pictorial models. In a certain sense every theory, whether employing direct pictorial concepts or not, is a model. Thus we may certainly consider the electromagnetic theory, in the form of the Maxwell field equations coupled with the appropriate definitions of the constructs they relate, as a model of the electrical and magnetic phenomena of our experience. However, to consider every theory a model is practically to drain the word model of its physical meaning as most physicists interpret it; therefore we shall concentrate on its pictorial significance. This leads to the conclusion that the general tendency in modern physics is to abandon such models in favor of formal systems cast in mathematical form. Quantum mechanics is an excellent example. Even in such theories it is ultimately necessary to deduce laws containing quantities that are epistemically defined (capable of experimental test). However, this does not mean that a pictorial model must be employed in order to provide epistemic significance for the constitutively defined constructs of the theory. In quantum mechanics one can obtain the observed frequencies of spectral lines that can be emitted or absorbed by a given atomic system directly from the calculated diagonalized energy matrix of the system by the use of the Bohr frequency condition (the relation giving the observed frequencies in terms of the energy values associated with the stationary states of the atom). No picture is necessary for this procedure.

Law

In Chapter One a physical law was defined as a symbolic shorthand statement descriptive of an observed routine of experience in what might be called the physical domain. As such it is tied to experience, and its scope is definitely limited to the domain it purports to describe. Since it is descriptive, it can only give a partial picture of experience, and there is no element of necessity in its application. In fact, the choice of the word law to symbolize this attempt to describe regularities of experience might be considered a questionable one, since "law" has such an imperative, judicial connotation in everyday language. It is probable that when it came into extensive use during the eighteenth century, there was considerable enthusiasm over an assumed necessary course of nature to which the well-known judicial and theological meanings of the word law might appropriately apply. This enthusiasm has long since evaporated in the face of the growing realization that the actual complexity of our experience severely limits the applicability of every physical law.

The situation takes on a different aspect when we consider the place of law in the context of physical theory. As we have already noted in this chapter, the fourth stage in the logical schema of a physical theory is the deduction of consequences from the postulates of the theory. We have called these consequences laws, but it is clear that they possess a different logical status from the descriptive laws we have discussed in Chapter One. When, for example, we assume Newton's second law ($F = ma$) to be applicable to the motion of any particle of mass m acted on by a net resultant force F, and ask ourselves what will be the motion of a particle subject to a constant force, we readily arrive at the consequence that in this motion the displacement must be a quadratic function of the time. This is a logical deduction from the assumption made, and there is no escaping it. If we decide to call this a law relating to motion, it is clear that logically the use of the term is different from that used earlier. The only justification for employing the word law in this case is that we can identify the logical deduction with the descriptive law for certain well-known motions, such as free fall near the surface of the earth. But this identification is the key to the whole purpose of a physical theory, which is to portray observed experience in a certain domain. The success of this process is measured by the extent to which the laws deduced from the postulates of the theory can be identified with already established descriptive laws or the ability of the theory to predict new laws that are in turn verified by experiment.

The relation between hypothesis and deduced law might seem at first glance to be straightforward enough. But closer inspection reveals an ambiguity that must be clarified, and this can best be done in terms of an example. The classical theory of thermodynamics, one of the most important theories in physics, is usually based on the so-called first and second laws of the conservation of energy and on the increase in entropy, respectively. From these hypotheses can be deduced a host of actual laws relating to the thermal behavior of matter, such as the Clausius-Clapeyron equation relating the latent heat of vaporization to the change in boiling point with pressure. In this context the first and second "laws" are not physical laws at all but postulates. However, it is clear that they might become laws in the strict sense if they in turn could be deduced from some more general theory. This turns out to be the case. For, by assuming that all matter is composed of component parts (atoms, molecules, etc.) and that the behavior of matter is controlled by the statistical co-operation of such component parts, one can deduce the first and second laws of thermodynamics and thus confer on them the logical status of actual descriptive laws. So we see that a postulate or collection of postulates in one theory can take on the aspect of law in another.

It is not surprising that there is a great temptation to try to transform postulates into laws by developing more general theories. Thus consider Newton's law of gravitation, which states that every particle of matter in the universe attracts every other particle with a force varying directly as the product of the masses and inversely as the square of the distance between the two particles, the coefficient of proportionality being the gravitational constant G. This is the basic hypothesis of Newton's theory of gravitation and from it follow the laws of planetary motion and other laws. Newton himself treated it as a postulate needing no further elucidation. But this has not prevented others from trying to deduce this principle from some more general theory. In George Louis LeSage's theory, for example, all so-called empty space is filled with small invisible particles moving in all directions with high speed. If any two large-scale material objects are placed in the field of such a system, it can be shown that due to the screening of each body by the other, a net force due to bombardment by the particles will tend to push the bodies toward each other and will vary inversely as the square of the distance between the centers of the bodies. This theory then has the law of gravitation as a logical deduction, and the latter no longer need be considered as a postulate. Unfortunately, though the conclusion of an attractive force obeying the inverse-square law attests to the plausibility of the theory, it fails to predict the correct order of magnitude of the force (the gravitational constant G). Or, put in slightly different form, in order to provide for the correct magnitude of the gravitational force, it is necessary to endow the LeSage particles with sufficient mass and velocity to heat the bodies they strike to such a high temperature that they will disintegrate. This is what Maxwell found on examining the theory. However, the challenge to produce a theory that will lead to the law of gravitation as a logical consequence still persists, and has led to other attempts, many of them, like Vilhelm Bjerknes', of a hydrodynamic character. The Einstein theory of general relativity is another illustration, though of a rather more subtle character.

The foregoing considerations emphasize the somewhat equivocal position of law and postulate in physical theory. Since a statement that is a postulate in one theory can be a deduced law in another, can such a postulate, which takes on the character of a deduced law, then become an experimental law? The answer would appear to be not necessarily Yes; indeed, in general it is No. Thus take again the case of Newton's law of gravitation. It cannot be an experimental law, since it does not describe a regularity of experience. Objection may immediately be raised that the Cavendish experiment, which has been repeated many times since it was first performed in 1797, is a direct verification of the law, and hence the law may be said to describe the experiment.[13] But this is not really the case. If we assume the law as a postulate, we can (given sufficient mathematical ability) compute the total attraction of two large-scale bodies for each other in terms of their masses, their geometrical orientation, and the gravitational constant G. Then by carrying out the appropriate measurement of the observed attraction, as Henry Cavendish did, we can determine the value of G needed to give agreement. The Cavendish experiment, in other words, is a method for determining G on the assumption that Newton's law holds. The expression for the actual attraction as deduced from Newton's law has the character of an experimental law for the particular experimental setup chosen, since it has the possibility of holding true independently of the masses of the bodies and the distance between their centers of gravitational action.

There are other illustrations of the equivocal position of law and postulate in physical theory, but it still appears that when a postulate takes on the character of a law, it will not necessarily be so in the descriptive, experimental sense. As another example, consider the various ways in which the theory of mechanics can be developed. One can begin with the Newtonian formulation in terms of the well-known three laws of motion, which are essentially a group of postulates. From these one can deduce the consequence (by defining energy, for example) that the time integral of the difference between the kinetic and potential energies of a system of particles (for the actual motion of the system between an initial and final configuration) shall have a stationary value with respect to all possible motions performed between these two configurations in the same time. This is the content of Hamilton's principle.[14] It is a direct deduction from the laws of motion in Newtonian form and hence has the character of a deduced law. On the other hand, one can proceed to develop mechanics as a theory by assuming Hamilton's principle as a postulate and can then deduce Newton's laws as consequences of this principle. This is a matter of taste. There is indeed some ground for the feeling that the Hamiltonian method is more general, in the sense that it imposes no such specific requirements on the definition of energy as the Newtonian method does. Thus by a somewhat arbitrary choice of energy dependence on velocity, one can deduce from Hamilton's principle the laws of motion of a set of particles in accordance with the principle of special relativity. These laws are different in character from Newton's in that the mass, considered an invariant in classical Newtonian mechanics, now becomes an explicit function of the velocity. But here again these laws are not laws in the experimental descriptive sense. They are differential equations that must be integrated under the imposition of definite boundary conditions in order to yield experimental laws.

Without boundary conditions physical laws would possess no utility. It will be recalled that the law of falling bodies can be deduced from the Newtonian second law ($F = ma$) by assuming that the force function F is constant and that the resulting second-order differential equation yields a solution in which the distance of the moving particle in question is a quadratic function of the time. However, before this solution can be compared with experiment, it must be made more specific. Actual numerical values must be assigned to the two arbitrary constants that appear in it. This may be done by assuming that the particle starts (at $t = 0$) at a definite position in the chosen reference system and with a definite velocity. These are the initial conditions of the motion; they may equally well be referred to as boundary conditions in time. With their use and the consequent evaluation of the constants in the solution, the latter becomes a law that can be tested experimentally.

There are also spatial boundary conditions that arise in describing the motion of continuous media. If, for example, we are treating the propagation of waves in a continuous elastic-fluid medium, the wave equation is the generalization of Newton's second law and serves as the basic postulate in the theory of wave motion. It is a partial differential equation and its solutions are general and in the form of arbitrary functions. They can only be made specific enough to compare with actual observable wave functions by the imposition of spatial boundary conditions (statements that the behavior of the medium over certain bounding surfaces has certain properties, for example, that the displacement may vanish there). Here again these conditions are indispensable if the relations deduced from the fundamental assumptions are to have specific physical meaning so as to be identified with experimental laws.

Appraisal

Physical theories are invented in the hope of achieving a better understanding of a portion of our human experience. Every theory has a logical structure that can readily be exposed and through which a firmer grasp of its meaning can be obtained. We must now face the question of what criteria to set up to appraise how well a given theory achieves its goal. There are several levels on which this can be approached. We can take a positivistic attitude and say simply that a theory is successful to the extent that, on the introduction of appropriate boundary conditions, the laws deduced from it can be identified with experimental laws. (In other words, the laws agree with experience.) This criterion becomes materially stronger if the theory predicts the existence of regularities in experience that have not hitherto been observed, and if the prediction is verified. This has been taken as a very powerful evidence of the value of a theory, and justly so.

Another criterion by which a theory may be judged is that of truth. This obviously implies a lot more than the positivistic appraisal, for it necessitates a decision as to the meaning of truth, and this is extremely difficult in any scientific context. Presumably those who believe that a theory should be judged at this level consider it the task of the physicist to discover the whole course of a world of phenomena that exist outside of and independent of the human observer. And having discovered these phenomena, he is to provide a unique explanation for them by means of a theory or set of theories. If and when a theory fits the facts, it will be said to be true, in the sense that the explanation will be final and not subject to upset by further discoveries. It is recognized by most adherents of this view that it is a highly idealistic one, in that no theory has as yet achieved the status of being completely true in this sense. Nevertheless, those who adhere to the criterion of truth believe that certain theories, such as the theory of mechanics, have shown themselves to be so plausible that all they need is further tinkering to make them good for all time. Those who hold this view show more confidence in our present methods of inventing theories than the positivists, and since this confidence may be an important ingredient in the psychological approach of the successful scientist, the view certainly has something to commend it. On the other hand, this same confidence may well be misleading to the layman when he learns of theories that have been discarded because they will not work and have been replaced by others that promise better. This has happened so often in the history of physics that it seems more plausible in the logical analysis of physical theories to discard the use of the notion of truth as it is commonly understood by philosophers and to fall back on the criterion of success, which can at any rate be understood and applied in a definite fashion.

The price that must be paid for the somewhat more positivistic attitude of the criterion of success is a willingness to attribute a purely tentative status to each physical theory. This inevitably puts stress on the essentially arbitrary character of physical theories. The positivistic approach appears to have other weaknesses, too. Philosophers have consistently and quite appropriately emphasized that simply because a particular theory leads to deductions agreeing with experience, we have no right to conclude that it necessarily constitutes the best explanation for the given domain of experience. There may well be several other theories that will do as good a job and perhaps even a better one. In fact, the history of physics confirms this with many instances of competing theories. On what basis, then, do we decide to give our allegiance to one theory rather than another?

The competition between the caloric and mechanical theories has already been discussed in this chapter. It is interesting to examine a case in which the attempt was made to provide a basis for choice between two theories by means of a decisive, or crucial, experiment. Such a situation arose in connection with the problem of the nature of light. During the eighteenth and part of the nineteenth centuries, two theories competed for attention in this field. One, usually associated with the name of Sir Isaac Newton, had as its basic assumption the postulate that light is propagated by means of small particles called corpuscles, which themselves cannot be seen or otherwise directly sensed, but which are emitted by luminous objects and travel in straight lines in a homogeneous medium. On entering the eye they produce the sensation of vision. They may also be reflected on striking opaque surfaces and refracted at the interface between two transparent media of different properties, such as air and glass.

Newton was able to show that the observed laws of light propagation are consistent with the corpuscular hypothesis; but the Dutch physicist Huygens, a contemporary of Newton, developed and championed another theory, the wave theory of light, in which the transmission of light takes place by means of wave motion through a medium in somewhat the same fashion as sound waves travel through air and other fluid media. As Huygens was able to show, this theory also accounts for the observed optical phenomena of reflection and refraction (including double refraction) and in addition those of diffraction (the bending of light around obstacles) and interference (evident in the colors of thin films, like soap bubbles). Newton was familiar with the wave theory but considered an insuperable obstacle to its acceptance the fact that it predicts diffraction, whereas light is observed to travel in straight lines and to cast more or less sharp shadows. (Newton was of course aware that the shadows cast by obstacles in the path of light beams are actually not completely sharp.)

The diffraction of light, which Francesco Grimaldi had already detected in 1666, is a much less obvious effect than the diffraction of sound. (We can readily hear around corners, but we cannot so readily see around them.) The diffraction of visible light is such a small-scale phenomenon that it takes rather special instrumentation to study it with any accuracy, though it can easily be detected qualitatively: merely by looking at a light source through the narrow slit between two fingers held close together, the usual diffraction bands can be seen. Of course such diffraction can also be accounted for on the basis of the corpuscular theory, though perhaps with somewhat more difficulty than on the basis of the wave theory, since the corpuscular theory demands that the particles of light be acted on by forces exerted by solid obstacles so as to deflect them slightly out of their original direction of propagation. Newton preferred this explanation, for he thought that wave diffraction should certainly be far larger in magnitude than the observed bending of light. He was also troubled by the fact that wave motion, as we know it in our experience, demands a medium. He could not visualize a medium elastic enough to permit vibrations such as light would have to correspond to and at the same time sufficiently tenuous to provide negligible resistance to the motion of the heavenly bodies. (Nineteenth-century physicists also found this a serious problem.) Newton was familiar as well with the colors of thin films, commonly explained in the wave theory by the concept of interference, whereby different wave trains are said to be able to superpose so as to augment their effects on the one hand or cancel each other out on the other. Newton, however, provided a corpuscular explanation of this phenomenon in terms of what he called fits of easy reflection and transmission.

The corpuscular theory maintained its popularity throughout the eighteenth century and even well into the nineteenth, in spite of the famous experiments of Thomas Young around 1800 on the interference of light and the elaborate mathematical theory of Augustin Fresnel around 1815. This state of affairs was by no means satisfactory to French physicists of that time, and it seemed obvious to many that it would be highly desirable to develop an experiment that would decide conclusively between the two theories. It had long been known that though both theories predict the law of refraction of light discovered experimentally by Willebrord Snell, the detailed assumptions necessary to derive the law are different in the two cases. Thus to obtain Snell's law, the Newtonian theory has to assume that the velocity of light of a given color is greater in a denser medium like glass or water than it is in a more rarified medium like air. On the other hand, to achieve the same agreement with experiment, the wave theory has to assume that the velocity in question is less in the denser than in the rarer medium. Although these conflicting assumptions would seem to have provided a basis for immediate experimental testing, up to the middle of the nineteenth century there was no way of measuring the velocity of light terrestrially. The accepted values were all based on astronomical data until about 1850, when two French scientists, Armand Hippolyte Louis Fizeau and Jean Bernard Léon Foucault, succeeded in providing the necessary terrestrial measurement with the rotating slotted disk and rotating mirror methods. Then the velocity of light in water was found to be less than that for light of the same color in air, and it was taken for granted that a crucial experiment had been performed, deciding conclusively in favor of the wave theory and against the corpuscular theory of Newton. This undoubtedly gave great encouragement to the view that the crucial experiment is a highly useful instrument in the appraisal of theories and therefore always to be sought out whenever two or more competing theories are under consideration.

But there are sobering second thoughts. Granted that the experiment on the velocity of light in water contradicted Newton's corpuscular theory, how can we be sure that it would contradict another or all other possible corpuscular theories of light? From the experiment itself it is impossible to decide just what element in Newton's theory fails to be verified. It is only the theory as a whole that is in disagreement with experience, leaving open the possibility that a revised corpuscular theory might pass the test. The French physicist and historian and philosopher of science Pierre Duhem has called attention to this difficulty with all so-called crucial experiments and has reached the conclusion that there are, strictly speaking, no real experiments of this character, since all theories are made up of more than a single construct and a single hypothesis, and no experiment is capable of singling out the weak part.[15] In fact, too literal interpretation of a crucial experiment can be dangerous for the progress of science by leading to the complete abandonment of a line of thought that conceivably might be useful later. And this is what has happened in the case of light. With the advent of quantum theory at the beginning of the twentieth century and with the discovery of the photoelectric effect and its various properties, it became necessary to resurrect the corpuscular theory, and we now have no hesitation in talking about photons, or particles of light.

The moral is that there seems to be no simple, clear-cut way in which to distinguish decisively between two physical theories explaining the same domain of experience. Yet as a practical matter, we have to do it and we do.

What other criteria for the acceptance of a physical theory can we find? An obvious criterion is provided by the principle of parsimony, or the famous "razor" of William of Occam: *Entia non sunt multiplicanda praeter necessitatem* ("Concepts that yield basic theories must not be multiplied more than is necessary"). Newton adopted this as the first of his "Rules of Reasoning in Philosophy" at the beginning of his "System of the World" (Book Three of the *Principia*), where he says: "We are to admit no more causes of natural things than such as are both true and sufficient to explain their appearances." He then goes on to amplify this statement: "To this purpose the philosophers say that Nature does nothing in vain, and more is in vain when less will serve; for Nature is pleased with simplicity, and affects not the pomp of superfluous causes."[16]

This has a fine sound, though Newton's confidence in the simplicity of Nature is a judgment probably not shared by the modern physicist. In the light of the present-day creation of experience in the realm of nuclear physics, it is a gratuitous assumption. Nevertheless, most physicists will tacitly agree, other things being equal, that that theory will be most acceptable which operates with the smallest number of independent ideas. The caloric theory, for instance, is in this respect at a disadvantage compared with the mechanical theory of heat, since the caloric theory has to introduce a new kind of substance, whereas the mechanical theory can operate with the ordinary matter of our experience. However, Occam's razor does not always work so effectively. How could it decide, for example, between the elastic-solid theory of light and the electromagnetic theory? Here it is very difficult to count constructs or postulates so as to demonstrate that one involves more than the other. Certainly the razor principle has no great trouble in disposing of purely *ad hoc* theories, in which different postulates have to be invented for each new application of the theory. But the sophistication of modern physics has gone far beyond this crude situation. The razor may indeed be of greater applicability to purely mathematical theories, where the constructs are relatively abstract and can be pinpointed, and where they are not identified with ordinary experience.

Much has been made of the concept of simplicity itself as a criterion for appraising the success of a physical theory. To some people one particular theoretical point of view appears simpler than another, even though the resulting deductions from the two can be equally well identified with experience. But if one probes deeply into the reason, it usually turns out to be a will-o'-the-wisp and dissolves into inexplicable preference or perhaps mere familiarity. We are apt to think of any idea as simple when we have thought long enough about it to become familiar with it. The person who has studied only algebra and geometry in mathematics is apt to think the concepts of the calculus mysterious and difficult and far from simple on first encountering them. But if such a person sticks to the subject he can learn it, and then it becomes simple—though higher analysis may then appear in another category again.

The effective use of a criterion of simplicity in the appraisal of physical theories will have to await a much deeper psychological study of the learning process. Meanwhile, physicists will continue to claim that some theoretical viewpoints are simpler than others and hence preferable. For example, Erwin Schrödinger, one of the originators of quantum mechanics, considered the matrix formulation of Werner Heisenberg so complicated as to be *abschrecklich* and obviously thought his own wave mechanical method of presentation was preferable. Since both mathematical methods lead to the same results, the choice here is really not between two theories, one of which is considered to be simpler than the other, but rather between two apparently different, but fundamentally equivalent, analytical ways of expressing the same physical content.

It is sometimes stated that physicists explain their preference for one theory over another in terms of the idea of elegance or beauty. Here we get pretty deeply into the realm of aesthetic value judgments, which are, to be sure, only a shade more difficult to deal with than the value judgments inherent in the postulates of a physical theory. The physicist is forever being confronted with the necessity of expressing preferences, of making choices, and often has a hard time explaining why he made the choice he did. In this he resembles the artist (or humanist in general) who has to face the same kind of situation. It is true that physicists of the eighteenth and nineteenth centuries felt themselves on surer ground with respect to their choices of fundamental attitudes and hypotheses than their successors of the twentieth. Many of these earlier physicists felt that the assumptions they were making were so naturally connected with, and suggested by, the phenomena to be described and understood that there was really little choice involved. When, for example, they decided to look upon heat as a substance, they were not impressed by any arbitrariness in this decision, since all the observed properties of heat in those days seemed to point more or less inevitably to the necessity for this assumption. In cases of this sort, it seemed somehow as if Nature were merely revealing and explaining herself, and man were merely following her lead. The whole development of physics and science in general, however, has cast increasing doubt on this simple interpretation. The twentieth-century physicist can no longer believe that things are what they seem. This makes his job steadily tougher as new experience accumulates, but it also challenges his imagination and gives him full scope for the development of new and sometimes bizarre ideas, to see how they work.

The price of the modern physicist's freedom in creating theories imaginatively is, of course, the tentative and somewhat arbitrary character of physical theorizing. For there is an increasing tendency, in the light of new scientific developments, to look upon the task of physics not so much as the discovery of experience already there and previously hidden from us, as the invention of new ideas in terms of which we may understand, or make ourselves think we understand, the complexities of new experience in the physical domain. The freedom to invent new constructs that has been grasped enthusiastically by many modern physicists was also grasped by earlier physicists. At this date it is difficult for us to put ourselves in the place of contemporaries of innovative physicists like Galileo, Huygens, and Maxwell. Their bold and daring ideas, which we now feel advanced immeasurably the understanding of experience, were in their own time undervalued by many and considered off the track of sensible theorizing. Hindsight is better than foresight, and it is likely that to our descendants the free inventions of today's quantum mechanicians and cosmologists will appear to have been satisfactory attempts at physical understanding, though probably by that later time outmoded in many respects.

In stressing the arbitrary nature of physical theorizing implied by free invention of constructs and postulates, we must be careful not to go too far. The physicist is certainly not privileged to introduce any idea he fancies independently of its relation to other concepts in the field; and as a matter of fact truly creative physicists introduce new concepts only after much soul-searching and endeavor to present as much justification as possible for new concepts in addition to the success of the deductions from them. The reason for this is clear: once a physicist believes he has hold of a good idea its arbitrariness ceases to impress him, and he begins to feel, in his bones, as it were, the essential correctness of his view. This is wholly appropriate, since unless a physicist takes his theories seriously, he cannot expect others to do so. At the same time he must subject his views to the critical examination of his colleagues and fellow workers in the same field. This serves to prevent, or at any rate mitigate, the rise of dogmatism as an obstacle to the advancement of science.

While the creator of a physical theory should have full freedom to use his imagination, no matter how esoteric and far removed from everyday experience the constructs and postulates are, it is obviously necessary that somewhere in the development of the theory the constructs or something derived from them be unambiguously identifiable with actual experimental operations. Moreover, this must be possible in more than an *ad hoc* fashion: the identification should cover the widest possible range of experience. For example, in the solution of Schrödinger's equation in quantum mechanics, the identification of the energy eigenvalues (when divided by Planck's constant h) with the spectral terms in the appropriate atom must hold for the application to all atoms, and as far as transitions from one energy state to another are concerned, it must be subject to the general selection rules provided by the theory. In fact, the more widely the identification can be stretched successfully, the greater the appeal of the theory. To take another example, when Planck's constant, first introduced in connection with the heat radiation law, also turned up in such different domains of experience as the photoelectric effect and the specific heats of gases, the plausibility of the quantum theory was considerably increased.

To sum up, the physicist who invents a new theory will use all the weapons in his arsenal to justify its plausibility and to encourage others to test it and try to enlarge the range of its application. Though he may be the first to admit that the workability of a theory in the pragmatic sense is no logical justification for assuming that it provides the final answer, yet he demands the right to continue to exploit a theory if he feels confident that it is an ingenious idea and if it helps him understand a certain domain of experience. In the last analysis we come back to faith in the value judgments of clever and imaginative people.

Chapter Three

SOME PHILOSOPHICAL PROBLEMS IN PHYSICS

Physics and Philosophy

All aspects of the logical analysis of physical theorizing involve philosophical questions. A few examples were mentioned in Chapter One, and these with others will be examined in some detail in this chapter. But first we ought to justify the assumption that philosophy has any role to play in physics at all. Many interpreters deny any serious connection between philosophy and a science like physics. They point out, for example, that philosophers seek to understand the world of experience in quite a different way from scientists. For one thing, philosophers rarely seem contented to abstract from the totality of experience small domains for intensive study. They wish, rather, to investigate ways of grappling with experience as a whole. They are much exercised over how we can know anything about experience and so have created a whole field of study called epistemology, or the theory of knowledge. When they ask what it means to say that we can know anything, they appear to be going well beyond the physicist's realm of interest, for the physicist takes it for granted that we can know and proceeds confidently from there.

Further, the philosopher raises questions about the nature of those aspects of human experience that the scientist says he wishes to describe and understand. Is this experience something wholly objective in character and independent of human observers, resulting from the existence of a real world, or is it something due entirely to the human observers themselves, something that exists only in their sense impressions? The philosopher thinks that such questions are necessary preliminaries to talking about understanding experience. The scientist on the other hand is apt to become impatient over such lucubrations and will often ask, "What difference does it make?"

This is not an appropriate place to discuss philosophy as a discipline. Interpretations of this differ considerably among professional philosophers themselves. But enough has been said to show that in spite of the physicist's impatience many of the searching questions of the philosopher do have relevance for physical theorizing. For example, it seems to be wiser for the physicist not to take the general ideas of space and time for granted as something that all intelligent people understand, but to look more closely at how these terms are used and at various possible modifications in their use in physics. We shall call this a philosophical problem in physics and shall pay some attention to it.

Similarly the philosopher's concern over the relation between events that he expresses in terms of the notion of cause can hardly fail to be considered seriously by the physicist, who is forever dealing with such relations. Hence we shall look at the meaning of causality from the standpoint of physics. This is another important philosophical problem in physics.[1]

Space and Time

Physical theory is inevitably dependent on primitive, intuitive notions—the ideas that the physicist takes for granted in order to begin his theory-building activity. The philosopher, however, considers it his duty to examine precisely these fundamental concepts; to him it is obvious that unless the foundations are clearly understood, the superstructure will be shaky indeed. Philosophers have therefore given a great deal of attention to these undefinables. Probably the most important of them all in physics are the notions of space and time.

The idea of space can be approached on three different levels: (1) private space, or the space of each individual observer; (2) public, or physical, space; and (3) conceptual, or mathematical, space. These distinctions will not meet with the general approval of philosophers, many of whom would prefer to develop a concept of space that is apparent to us all in an absolute sense, whether we are physicists or not. This view certainly deserves careful consideration, and we shall not overlook it; for the moment, however, we must emphasize what everyone will admit, that as each one of us looks out on the world of experience, he feels an irresistible urge to find some order in it. As we have already observed in Chapter One, this is part of the urge toward scientific description in the general sense. But order involves in the first place recognizing a distinct object at each distinct moment of time, or as it may also be put, telling things apart at any given instant. This individual recognition is what we shall mean by private space, and following the great German philosopher, mathematician, and scientist Leibniz, we shall define it as "the order of coexistent sense impressions." This definition creates other difficulties, since the word coexistent cries out for independent definition. However, most physicists will agree that they think they understand it. Human perception indeed is not completely subsumed by this type of order. It demands that the order of successive events at the same place be reckoned with, and this involves the concept of time, also a primitive notion. Like space, time has initially a purely private aspect. It is the property of the individual sentient being and is connected with his need to estimate duration and his ability to do so in terms of changes taking place in his own body.

Even when we talk about space in its most private aspect, as a psychological concept, it is therefore necessary to introduce the idea of time. Similarly it is impossible to think logically of time even in the duration sense without invoking the notion of space. For we know that anything we call a clock (whether it be the earth, a small-scale mechanical device, or something going on in different parts of the human body) involves the motion of an object or objects in space. This interdependence of the two concepts strongly suggests that the analysis of the order of sense impressions into the two categories space and time, while convenient in some respects for the affairs of everyday life, is fundamentally artificial as far as scientific description is concerned. A synthesis of the two notions into a single representation of events would appear to be more logical. As a matter of fact, this is precisely what the theory of relativity has brought about.

Before we go into such elaborate matters, however, we ought to revert briefly to what we have called private space, or what might be more appropriately termed psychological space. We have really no right to refer here to "space" in the singular. Actually there are private spaces, and indeed one for each mode of sensation: visual, auditory, tactile, motor, olfactory, and gustatory. When what is seen is arranged in a certain order, this order may properly be called visual space. A blind man arranges his experiences in terms of the sounds that reach him and often develops a very acute spatial sense. A person both blind and deaf still manages to introduce order in his surroundings by means of touch and motion—by feeling objects he can estimate their size and shape, and can pace off distances between them; we say he has a motor sense. The spaces based on the senses of smell and taste are obviously much more limited in extent and variety than the others, but they are nevertheless genuine. It is a part of the education of a child to integrate these separate private sensory spaces into a single psychological space in order to simplify the interpretation of experience; he learns, for instance, to use the evidence of tactile and motor space to convince himself that a coin that looks quite different in size and shape at different distances and orientations is really the same coin. Similarly sounds and sights are brought into line to avoid illusions and deceptions.

This kind of sensory synthesis does not result in a space that is uniform for all observers, since sensory perception is subject to individual differences: Some people see and hear much better than others. Even when these differences have been averaged out and accounted for, psychological space is saddled with properties that make it inadequate as a matrix, or container, of scientifically described experience. It is finite in extent and is also nonisotropic (different in different directions) and nonhomogeneous (different in different parts). Moreover, it is really discontinuous, as the blind spot in the eye amply demonstrates, and has no well-defined number of dimensions.

So that people can communicate with each other, it is essential to replace the individual, psychological spaces of each observer with an abstraction from them all, with a kind of idealization in which certain properties are attached to space for the sake of convenience. The mind constructs the idea of a physical object whose relation to the observer is determined by the properties of an ideal space called physical space. The properties of physical space must be such that measurements (see Chapter One) can be carried out meaningfully. If, for example, we make a scale by introducing marks on a rigid rod, it is necessary that the interval between any two marks shall remain the same, no matter in what direction we turn the rod or how we move the rod from one place to another. In other words, we assume that the rod remains rigid independently of position and motion. This is the postulate of free mobility in space. It is an assumed property of physical objects to assure the possibility of consistent measurements. For convenience, however, it is customary to associate this property with physical space itself and to assume that this space is isotropic and homogeneous in contrast to private, psychological space.

By experiments on physical objects it was learned a long time ago that there are certain relations connecting such objects, and these relations somehow seem independent of the objects themselves, as long as the objects are rigid. The concept of a closed figure formed by straight rigid rods emerges, and certain properties connected with the angles made by the various pairs of rods come to light through measurement. If the figure is a plane triangle, for example, whatever its size, the sum of the three interior angles turns out to be so close to two right angles as to lead to the conclusion that a property of space itself has been discovered that has essentially nothing to do with the physical nature of the rods themselves. Here arises the concept of geometrical space, illustrating a physical theory of precisely the kind discussed in the two previous chapters. Such primitive undefined notions as "point," "line," "straight line," and "plane" are introduced. In the Euclidean formulation these are indeed given definitions like "A point is that which has no parts," but these definitions amount to saying that we know from ordinary crude experience what we are talking about when the word point is used. Once having settled on these fundamental ideas, we can proceed to construct more precise concepts—figures like triangles, polygons, and circles, for instance. Then we begin to introduce hypothetical statements in the form of postulates. In Euclidean geometry these are called axioms and postulates; the distinction presumably was made by Euclid because he thought that certain of the statements in question were self-evident, while the others were only plausible assumptions. Examples of axioms are: "Things equal to the same thing are equal to each other" and "The whole is greater than any of its parts." Illustrations of postulates are: "All right angles are equal"; "It is possible to draw a circle with given center and through a given point"; and the famous parallel postulate, "If two straight lines in a plane meet another straight line in the plane so that the sum of the interior angles on the same side of the latter straight line is less than two right angles, the two straight lines will meet on that side of the latter straight line." This formulation is called the parallel postulate because Euclid found it necessary to assume it in order to prove that through a point not on a given straight line there is one, and only one, line parallel to the given line.

From these concepts and hypotheses the theorems of geometry were derived. These theorems are the deduced laws of geometry treated as a physical theory. One of these laws, for example, states that the sum of the interior angles of a plane triangle is two right angles. This, like the physical laws that have been explained through the agency of a theory, is both an experimental law and a deduced law. We can actually take a protractor and measure the interior angles of a large number of plane triangles of various sizes and shapes, and verify the validity of the law within the limits of the precision of the instrument. The other theorems of geometry have the same logical and experimental status.

Why is geometry not actually considered to be a physical theory? It is claimed as a branch of mathematics, presumably because of the early conviction of mathematicians that the whole intellectual content of geometry —concepts, hypotheses, and theorems—applies not merely to the properties of physical objects but also to the space that is the matrix of physical phenomena (that space which is the basic order necessary to conceive the occurrence of such phenomena). According to this interpretation geometry is a description of the properties of space needed for physics. And when the word geometry is used here, for most practical purposes it means the geometry of Euclid, which we all learn in secondary school. In fact up to some 130 years ago it was believed that Euclid's was the only possible geometry and that its use was forced on the mind because without it one could not function in physics. Kant made Euclidean geometry a part of his philosophy, treating its axioms and postulates as a priori synthetic judgments, necessary conditions for obtaining knowledge from human experience.

Ever since the invention of non-Euclidean geometries, it has been realized that Kant was mistaken. After many attempts to prove the parallel postulate as a theorem resulting from the other postulates of Euclid (usually by the *reductio ad absurdum* method of denying the postulate and trying to show that a contradiction ensues), mathematicians concluded that it is perfectly possible to invent consistent geometries by denying the parallel postulate. Farkas Bolyai, in 1832, and Nikolai Ivanovich Lobachevski, in 1835, invented a non-Euclidean geometry by assuming in effect that through a point not on a given straight line, it is possible to draw an infinite number of parallel straight lines. In 1854 Georg Friedrich Bernard Riemann suggested the invention of another type of non-Euclidean geometry in which he assumed in effect that through a point not on a given straight line it is possible to draw no straight line parallel to the given line. Both these geometries have been developed in completely consistent fashion, and though their theorems are different from those of Euclidean geometry, no one has discovered a contradiction in them. In Riemann's geometry the sum of the interior angles of a triangle is greater than two right angles; in the geometry of Bolyai and Lobachevski, the corresponding sum is less than two right angles. The two-dimensional analogue of Riemann's three-dimensional geometry is spherical Euclidean geometry. The corresponding Lobachevski-Bolyai analogue is the geometry on the surface of a tractroid.[2] In both geometries free mobility is assured. (In the space described by both geometries, physical objects may be moved from place to place without suffering change in their dimensions.) This property is shared with the space of Euclidean geometry. Mathematically it is expressed by saying that all three spaces are characterized by constant curvature. On the other hand, the two non-Euclidean spaces differ from Euclidean space in that in non-Euclidean spaces it is impossible to have similar figures of different dimensions. For example, to consider again the analogue of spherical geometry, there are no similar triangles on a sphere. Euclidean space is called flat or homaloidal, while Riemannian and Lobachevskian spaces are curved or nonhomaloidal.

Now it is obvious that the theorems or laws of non-Euclidean geometry do not agree with our terrestrial experience within the margin of normal experimental error, and it might therefore be thought that they are merely a mathematical curiosity and of no significance for physics. However, we are interested not only in the space of our immediate surroundings but also in astronomical space. The legitimate question arises of how we can be sure, when we are dealing with distances of the order of those between the sun and planets of the solar system or those between the stars, that space still continues to possess its Euclidean properties. In this case we no longer have the opportunity to perform experiments such as those we can perform on earth with rigid measuring sticks; rather we are forced to measure with light rays. This may indeed be changed if space travel on a large scale becomes feasible. But at the present time the results will depend on the theory we adopt for light propagation. On the surface of the earth we find it convenient and satisfactory to assume that light travels in straight lines through a homogeneous medium. (The use of surveyor's instruments is based on this assumption.) It is a questionable extrapolation to assume that the same holds for intergalactic space. Tests have been carried out for the justification of the assumption that Euclidean geometry applies to the far reaches of astronomical space. For example, measures of the parallax of stars have been studied to see whether a negative parallax could be detected in any case. This would mean that if a star were observed at two times six months apart, the sum of the angles made by the line of sight from earth to star and the diameter of the earth's orbit would be found to be greater than 180°. If such a situation had ever emerged and been verified for other and more distant stars, a reasonable conclusion might be that over great distances space is really non-Euclidean. As Poincaré emphasized, however, we should not really be justified in maintaining this conclusion, since it would be based on the assumption that light rays follow straight-line paths, whereas they might actually be curved slightly over very great distances.[3] His conclusion was that we shall never know by experiment what geometric space is most satisfactory for physics but must simply settle on the one we deem most convenient. This conclusion has, however, been altered considerably by the advent of relativity.

As we have already suggested, the interest of physics in space is really connected with the type of change in physical experience that we call motion. No matter what view of space we hold, we must face the question of whether the properties of space are the same for moving objects as for stationary ones. This means that we are not justified in discussing space as a mode of ordering physical experience without introducing the concept of time.

As with space, we can treat the idea of time on several levels of sophistication. The notion of duration appears to be a private property of the individual, and we are all familiar with it. We talk of the passage of time in terms of bodily sensations. We also observe changes in our immediate environment and finally decide as a matter of public convenience to adopt these as a means of measuring time in a standard fashion that is acceptable for the uses of all people. This leads to the concept of public, or physical, time, or as we may quite appropriately term it, clock time. With a strong sense of the inadequacy of clock time because of its variability from one piece of mechanism to another, Newton sought to establish the concept on an absolute basis with his famous definition: "Absolute, true and mathematical time or duration flows evenly and equably from its own nature and independent of anything external; relative, apparent and common time is some measure of duration by means of motion (as by the motion of a clock) which is commonly used instead of true time."[4]

Newton's definition, of course, is circular, but all definitions are ultimately circular, since every word has to be defined or talked about in terms of other words. The modern interpretation of Newton's definition is that time effectively appears in physical theory as a parameter t, which can vary continuously and take on all the values of the real number continuum. Thus time serves as an independent variable in terms of which other variables characterizing physical objects can be expressed by suitable functional relations. These other dependent variables might, for example, be the displacement and velocity of a particle, the temperature of a liquid bath, or the intensity of an electric field. The whole purpose in setting up such functional relations of physical properties in terms of a parameter like t is to have a simple analytical way of representing change in physical experience. The fundamental mathematical equations that physical theories set up as expressions for the theoretical hypotheses are usually differential equations. When these are solved subject to the conditions conformable to the given physical situation, they yield such explicit expressions for the physical properties in terms of t as the distance of a particle from a chosen origin as a function of the time. Now in order that this result shall have any meaning as a description of experience, we must give some operational significance to the parameter t; in other words we must measure it. This necessitates going back to public time and inventing a suitable clock, which, after all, is only an arbitrary piece of mechanism in which a pointer moves over a scale.

In a sense this lets the cat out of the bag as far as the real meaning of time in physics is concerned. For what we are actually doing when we allow a clock to provide values of t is to describe one physical system in terms of another. That is, we introduce a one-to-one correspondence between the system whose changing behavior we are trying to describe and the arbitrary physical system we call the clock. From this point of view there is no reason why we should not dispense entirely with time in physics merely by deciding to describe the whole of experience in terms of the comparison of physical systems.

Two physical systems that may be compared in this fashion are a particle attached to the end of a spring, the other end of which is fastened to a rigid support, and a freely falling particle. Or, the freely falling particle can be replaced by the bob of a simple pendulum. Matters are arranged so that the motions are in the focal plane of a motion-picture camera, and the two motions are photographed. The exposed and developed film will present a one-to-one correspondence between the motions: that is, to each position of the one particle a position of the other will correspond. If we call the displacement of one particle from an arbitrarily chosen origin x and that of the other y, we can then plot y as a function of x along perpendicular axes in a plane and obtain a curve that will be the geometrical representation of the relation between the two motions. (Note that this representation will be independent of the movement of the motion-picture film.)

Other pairs of systems can be compared in similar fashion. However, anyone who tries this scheme will undoubtedly soon reach the conclusion that it cannot provide a very useful way of describing the motions of physical systems unless we decide to choose a certain standard system with which to compare all others. For practical purposes this system will be one in which the motion repeats itself in cyclic fashion so as to be confined to a relatively small portion of space. This system then becomes a clock, and the co-ordinate attached to its scale will be called t. For further convenience it is desirable for this clock to be put in correspondence (be synchronized) with all other such systems designed to be used as clocks. Finally, it is almost inevitable that the behavior of the clock will be so adjusted as to bear some relation to the psychological sense of duration in the human being. Of course this is arbitrary, but it reflects the fact that physics as a branch of science is created by and for human beings.

Measurement of time by any of the physical systems called clocks involves the comparison between two physical systems as they change their position in space. Because of the injection of the purely psychological idea of time into the construction of clocks, we tend to overlook the spatial nature of time's actual measurement. This raises the fundamental question of why we should wish to describe physical experience in terms of the relation between physical systems as indicated. Why should we not look primarily, if not exclusively, at those aspects of experience that are independent of such interrelationships? This question in turn may be traced back to antiquity—to two different philosophical viewpoints, represented respectively by Heraclitus of Ephesus and Parmenides of Elea, both of whom flourished in the sixth century B.C. Heraclitus, impressed by the evident flux of experience, saw change as the most important thing about it and said: "Nothing ever is, everything is becoming"; "All things flow"; "You can never step into the same river twice." On the other hand, Parmenides believed that change is an illusion, that nothing really changes, and that the important things in experience are the invariants. He wished to stress those aspects of experience that maintain their identity in the midst of apparent change. These aspects are morphological in character, referring to forms and shapes, as distinct from changes.

An illustration from astronomy will make clearer the distinction between the two points of view. To describe and understand the observed behavior of the planets in the sky, we can emphasize the "change" aspect of the situation and try to express the motion of the bodies in question in terms of the earth as a standard physical reference system. This implies finding the position of a planet as a function of the parameter t, where t is the expression for the position of the earth, otherwise known as the time. This might be called the Heraclitean approach. Astronomers have to use it to prepare the ephemeris, or table of positions of the planets, but it is difficult to carry out mathematically and involves much calculation. It is much simpler to follow the Parmenidean idea, in which case the expression for

Newton's second law of motion (the net external gravitational force of the sun on the planet considered effectively as a particle equals the product of the mass of the planet and its acceleration) takes the form of a vector differential equation of the second order. This relates the second derivative of the position vector of the planet (still considered effectively as a particle) with respect to the time t, to the vector itself through the inverse-square law of gravitation, where the position vector is the vector connecting the sun with the planet. This is a problem in what is called central field motion, since the force attracting the planet to the sun is always directed from the planet to the sun. From this fact, even without solving the equation of motion for the position vector as a function of t, we can draw certain general conclusions. Among them is the law that in all central field motions the rate at which the position vector traces out area in the plane of the motion as the planet moves around the sun is constant. (This law of areas is also known as Kepler's second law of planetary motion.) By using this law and the equation of motion one can, with appropriate mathematical manipulation, eliminate the parameter t from the problem and effectively transform the equation of motion into one that relates the magnitude of the position vector (the radius vector r) with the angle θ that the line joining the planet to the sun makes with some arbitrarily chosen axis in the plane of the motion. This differential equation can be solved for r as a function of θ, and the resulting equation is the analytical expression for the curve that represents the actual path of the planet. For appropriate initial conditions the orbit becomes an ellipse (Kepler's first law). Here we learn something physically significant about planetary motion without any reference to time, obtaining what is called a morphological result. The result is also holistic in that it represents the motion of the particle in question as a whole and indicates that there is a certain pattern or shape of motion that is important in the world of our experience and challenges us to account for it.

Physics is of course full of such relations, among them the laws of statics in mechanics, according to which, for example, a flexible string, suspended from any two points and acted on by gravity only, assumes the shape of a catenary. Another is Archimedes' principle in fluid mechanics, which is again a morphological result having nothing to say about time. In fact all statements about systems in equilibrium are of this character. They are essentially spatial rather than temporal in their nature, as are the equations of state of substances subject to the laws of thermodynamics.

Laszlo Tisza has stressed that quantum mechanics provides another good illustration of the morphological approach.[5] In what we might more appropriately call quantum statics, the aim is to find time-independent forms of natural phenomena on the atomic level—not sequences of events. One of the fundamental problems of quantum statics is the calculation of the "states" of an atomic system, that is, the various energy values it is allowed to have. These values are independent of how the atom got into the states in question. The Bohr frequency condition—which relates these energy eigenvalues, as they are called, to the possible emission and absorption frequencies of the atom—is also a morphological equation.

This sort of representation can be imposed on any dynamic phenomenon. Consider the introduction of phase space in statistical mechanics. This is the space any point of which has co-ordinates representing respectively the positions and momenta of a set of particles moving in ordinary three-dimensional space. A single path in phase space thus represents all the events constituting the motion of the set of particles. Take, for example, a simple harmonic oscillator of mass m and stiffness k. If we denote the displacement of the oscillating particle from its equilibrium position by q, the potential energy of the particle is $\frac{1}{2}kq^2$. The momentum p is $m\dot{q}$, where $\dot{q}$ is the instantaneous velocity of the particle. (The dot above q denotes differentiation with respect to t.) The kinetic energy $\frac{1}{2}m\dot{q}^2$ then becomes $p^2/2m$, and the total energy E is expressed in the equation:

$$E = p^2/2m + kq^2/2 . \qquad \text{(Equation 1)}$$

The system is a conservative one: in any given motion the energy remains constant in time. If we plot the preceding energy equation (for constant E) in the (p, q) plane we get an ellipse, whose semimajor axis is equal to $\sqrt{2E/k}$ and whose semiminor axis is equal to $\sqrt{2Em}$. As the actual particle goes through its oscillatory motion in ordinary physical space, a point may be thought of as moving around the phase ellipse, and a complete circuit of such a point corresponds to a complete cycle of the actual motion.[6] The motion is thus in a sense "spatialized" by the phase curve. This notion of phase space was utilized very extensively by J. Willard Gibbs in his presentation of statistical mechanics.[7] It provides a morphological approach to the dynamical behavior of complicated systems with many degrees of freedom—systems in which the sequence in time becomes hopeless to follow but in which averages over phase space can be identified with some measurable properties of the system. The success of the method of statistical mechanics depends on the extent to which the basic postulate (that the time average of an actual measurable property of a system can be identified with the average of this quantity over phase space) agrees with experiment.

There is an important observation to make about the morphological method of representation. In practically all the relations that result from this method quantities occur that demand the use of the parameter t in their definition. Thus in phase space we cannot define the momentum p except in terms of the time, since it is definitely connected with the velocity $\dot{q}$. Moreover, the energy E in Equation 1 exhibits a similar dimensional dependence on t. In the famous Bohr frequency condition both the energy levels and the frequency involve reference to time. Even in the case of the elliptical planetary orbit—which we took as an illustration of the elimination of the time from planetary motion with the consequent holistic representation of the motion in terms of its spatial shape—we must admit that the dimensions of the ellipse (the semimajor axis, for example) depend on the period of motion of the planet in its revolution around the sun. So t enters inevitably as soon as we try to use the spatial representation for any purpose other than merely to contemplate it as involving one particular shape rather than another. If time as represented by the parameter t is indeed a devil, it then appears that in the spatialization process we have merely hidden him and not exorcized him completely. The spatialization process may well be a handy expedient, but it does not really circumvent our essential preoccupation with change when we describe physical systems. But change involves the comparison of a given system in its various phases with a standard system, and hence t creeps in whether we will or no. This of course does not prevent us from concentrating attention on such aspects of our experience with physical systems as momentum and energy, which under certain conditions remain constant while other aspects change. We cannot fail to be interested in conservation principles. But it is probably significant that even conserved quantities involve in their very definition the notion of change and hence the parameter t.

From the way in which we have introduced the concept of time in physics, it should be clear that anything like an absolute time in terms of which all physical systems should be described is a chimera. The time t we use depends in any case on the system we decide to use as a clock. The philosopher will object that the indefiniteness implied in this statement is not so great as it might at first appear to be, since if we wish to preserve any meaning in the description of change in physical systems, we must use compatible clocks: we must see that the various clocks we use all run the same way or, if they do not, we must be able to compensate for their differences. This would appear not to give much trouble, since we would merely have to keep a whole set of clocks together at one place and observe their behavior, adjusting them for synchronism.

It turns out that things are not quite so simple as soon as we need to compare systems that are not at the same point of space. We are then confronted with the problem of simultaneity. If it is now 10 P.M. where we are on the surface of the earth, what is the time on the star Sirius? Even the believer in absolute time may take pause at this question. He will certainly not wish to answer, "10 P.M.," for he knows that when it is 10 P.M. where he is, it is not 10 P.M. at a point halfway around the earth. But when it comes to assigning a time to the simultaneous event "now" on Sirius, he will be at a loss. Of course he may say that if "now" means when I look at my watch and find it reads 10 P.M., my fellow on Sirius can very well read 10 P.M. on his watch "now," if his watch is synchronized with mine. The big problem concerns the "now." How do we know when the man on Sirius should look at his watch to correspond to my "now?" The obvious answer in the case of two persons within viewing distance of each other is to say that each observes the other looking and this fixes "now." Here we have tacitly assumed a signal of some sort passing from one to the other. If such signals were transmitted without any delay, the fixation of "now" could be carried out, but all experimental evidence seems to confirm that this is true of no signals in our experience: even light travels at a finite velocity. It follows that the definition is an arbitrary affair, in the choice of which we are guided principally by the desire for mathematical simplicity (itself a somewhat illusory criterion, which we shall examine later in this chapter) plus a few logical considerations.[8]

The simplest way of arriving at the definition of simultaneity that Albert Einstein introduced is to follow his own method. To do so, we must suppose that every point in the universe where the events that will engage our attention are to take place has a clock attached to it. The problem is to synchronize these clocks. They are assumed to be alike when an observer examines them all at one place. Consider two points P_1 and P_2 with their respective clocks at rest there. Let a light signal leave P_1 at time t_1 as registered on the clock at P_1 and arrive at P_2 at time t_2 as registered on the clock at P_2. It is then reflected back to P_1 and arrives at time t_1' as indicated on the clock at P_1. Einstein said that the two clocks run in synchronism if

$$t_2 - t_1 = t_1' - t_2 ,$$

or

$$t_2 = 1/2 \cdot (t_1 + t_1') .$$ (Equation 2)

Expressed in words, this means that the time the signal reached P_2 as recorded on the clock at P_2 is taken to be the arithmetical mean of the times of the signal's emission at P_1 and return to P_1. Once clocks have been synchronized in this fashion, the time of an event taking place at any point in space is then simply the reading on the clock at the point where the event took place.

Einstein's definition of simultaneity leads us at once to the question of how, if we try to assign times to the same event in two reference systems moving with respect to each other with constant velocity, the assigned times will differ, if at all. With this question we move directly into the domain of the theory of relativity and to the fact that, long before Einstein, Newton knew that it is impossible to detect the motion of a ship going through the water at constant speed by any mechanical experiments performed on the ship. Thus, if a person on shipboard throws a ball straight up in the air, it returns via the same route and does not fall at a point displaced from its original projection, contrary to views which have been expressed at various times. To be sure, this ignores the rotation of the earth and the Coriolis acceleration, which produces an easterly deflection of a falling object from the plumb line. But this is a small effect and can be neglected in the present considerations. We may assume that the ship is traveling along a perfectly flat motionless liquid surface. The reason that mechanical experiments like this one cannot demonstrate the motion at constant speed is simply that the equations of motion (from Newton's second law) are invariant in form with respect to a transformation of coordinates to a reference system moving at constant velocity. Such a transformation has the mathematical form

$$x' = x - vt ,$$ (Equation 3)

where x is the distance of a moving object from the origin of a fixed reference system at the time t measured on a clock fixed to this reference system. Then x' is the distance of the same object from the origin of the reference system moving at constant velocity v with respect to the first system. If we introduce this transformation into the second-order differential equation that is the law of motion of the object, the equation will be found to have the same mathematical form in x' that it had in x, which means that there is no way to distinguish which reference system the particle is moving with respect to; nor is there any way to detect the absolute motion of the two reference systems. It must be pointed out, of course, that the transformation (Equation 3) implies that the time t' as shown on a clock attached to the moving reference system is the same as the time t shown on an originally synchronized clock attached to the fixed system. (Newton thought this was obvious, for he felt that in mechanics he was always dealing with absolute time, which had nothing to do with the motion of clocks.)

The inability to detect by any mechanical means the absolute motion of two systems moving with respect to each other at constant velocity, which may be called the Newtonian principle of relativity, at once stimulates another question, that of whether there is *any* physical means for detecting the absolute motion of two reference systems moving at constant velocity with respect to each other. The immediate answer seems to be that there is, of course, an optical means. All we have to do is to look about us, and we see relative motion. But this suggests the question of whether the optical method will enable us to tell, in the case of relative motion, which system is really moving in an absolute sense. Thus for the ship on the water, how can we tell whether the ship is moving and the water surface standing still or whether the ship is standing still and the water surface is moving away in the opposite direction? As far as our sight is concerned, either alternative is acceptable. Put in these terms the problem may be puzzling, but it hardly seems of much consequence. However, something like this was a great puzzle to the nineteenth-century physicists, who were much concerned whether to attach meaning to the statement that the earth moves through space in some sense fundamentally different from its motion relative to the sun and the planets in the solar system. Physicists at that time believed that light travels in a stationary medium surrounding all actual physical objects, and they thought that by some optical means it ought to be possible to detect the motion of an object like the earth through this medium. The famous Michelson-Morley experiment was an attempt to test this by the use of very high-precision optical equipment. There has been a certain amount of controversy over the results of this experiment as repeated over many years from the original version in 1881 until well into the twentieth century. Nevertheless, the general verdict now is that this experiment, like other more purely electromagnetic ones, yields an essentially negative result and that it is impossible by optical or indeed by any other physical means to detect the absolute motion of the earth (its motion relative to the light-bearing ether of space—the medium through which light reaches us from the stars, for example).

Einstein erected his theory of restricted, or special, relativity on the fundamental assumption that it is impossible by any physical means whatever to detect the motion of an inertial system (an inertial system being one that moves with constant velocity with respect to all other such systems). Put in another way this means that all physical laws are of the same form in all such inertial systems, for if the forms differed, we should have a way of distinguishing one system from another and hence a way of deciding that some really move in absolute fashion and others do not.

The really important thing about Einstein's principle of relativity is the nature of the transformation equations that must be used to describe events occurring in one inertial system in terms of the co-ordinates appropriate to another such system and to the time as indicated on the clocks fixed in the other system. Contrary to what at first might be supposed, the transformation is not that of Equation 3, though it reduces to this if the velocity v is very small compared with c (the velocity of light in free space). Much more important from the standpoint of our ideas about the nature of time, the original and naïvely plausible Newtonian assumption that times will be measured the same way in any two inertial systems (that t' will equal t) turns out to be untenable.

If this identity were to prevail, the equations of the electromagnetic field, which govern the propagation of light and other electromagnetic radiation, would not remain invariant in form for all inertial systems, and the fundamental idea of relativity would be violated. In order to preserve this idea of the fundamental indistinguishability of all inertial systems by any physical experiments whatever, we must give up the notion that there is a kind of absolute time that applies to all inertial systems. Instead, if we denote by t the time of a given event in one system, we must denote the time t' to be assigned to the same event on clocks associated with the other system by

$$t' = \frac{t - vx/c^2}{\sqrt{1 - v^2/c^2}}, \qquad\qquad \text{(Equation 4)}$$

and indeed replace Equation 3 by the equation

$$x' = \frac{x - vt}{\sqrt{1 - v^2/c^2}}. \qquad\qquad \text{(Equation 3')}$$

In doing so, we are assuming for mathematical convenience that the inertial systems are systems of rectangular co-ordinates moving along their common x axes with velocity v. Equation 3′ and Equation 4 are called the Lorentz-Einstein transformation equations of restricted or special relativity.

There has been a good deal of misunderstanding of the significance of these transformation equations, particularly from the standpoint of the philosophical meaning of the concepts of space and time. It has been asked, for example, whether, if time is really different in the primed reference system from what it is in the unprimed system, this does not contradict the intuitive feeling that after all there can be only one real time in human experience. The only answer is that in order to write equations that will hold in the primed reference system referring to events described by x and t in the unprimed system, x' and t' as given in Equation 3′ and Equation 4 must be used in place of x and t. Otherwise the result will permit a distinction between the two systems in contradiction to the fundamental assumptions of the theory of relativity, and this in turn will lead to results in disagreement with precise experiments. It must be emphasized that this does not mean mere disagreement with such null experiments as the Michelson-Morley. Rather, the theory of relativity, through the transformation equations (Equation 3′ and Equation 4), leads logically to striking positive conclusions that can be tested by experiment. One such conclusion, for example, is that the mass of a fundamental particle like an electron varies with its velocity v with respect to the reference system fixed in the laboratory. This variation is expressed by the following formula:

$$m = \frac{m_0}{\sqrt{1 - v^2/c^2}}, \qquad\qquad \text{(Equation 5)}$$

where m_0 is the mass for the particle at very low velocities, or strictly at rest ($v = 0$). This formula has been completely verified by experiment.

One can take the attitude that relativity has really nothing to do with the philosophical interpretation of time as a category of sense perception, but merely instructs us how to use the parameter t in our equations so as to satisfy certain assumptions that form the basis of an interesting and plausible physical theory. This theory in turn predicts results of significance in the behavior of the small, fast-moving particles that form the stock in trade of that branch of physics called atomic. But the fact that the conclusions of Einstein's relativity theory play no significant role in macroscopic physics (because the particle velocities encountered in such physics are all very small compared with c) is irrelevant to the main issue, which is the right of the theory to be tested on its own ground in the same fashion as any other physical theory.

From the philosophical point of view, perhaps the most striking predictions of the special theory of relativity are the contraction in length of a moving object in the direction of motion and the contraction in time intervals in a moving clock. If l is the length of a rod measured in one reference system, then to be consistent with relativity the length l' of this rod in the reference system moving with velocity v with respect to the first is

$$l' = l \sqrt{1 - v^2/c^2} , \qquad \text{(Equation 6)}$$

it being assumed that the axis of the rod lies in the direction of motion. This is the famous Fitzgerald contraction, originally assumed in order to account for the presumptive null effect of the Michelson-Morley experiment. Similarly if Δt is the interval of time between two events as measured by a clock fixed to a given reference system, then in the reference system moving with respect to the first with velocity v, the interval between the same two events becomes $(\Delta t)'$, where

$$(\Delta t)' = \Delta t \sqrt{1 - v^2/c^2} . \qquad \text{(Equation 7)}$$

This is the basis for the statement that according to the theory of special relativity a moving clock runs slow. It has produced much philosophical debate, mostly centered again on the theme that there is a real time that has nothing to do with clocks in moving reference systems. Obviously, the results of relativity will have little appeal to a person who has preconceived, strongly held views of absolute time and space. Nevertheless, the results of the theory of relativity, insofar as they have been tested, are in agreement with experiment, and therefore physicists will continue to work with x and t in their theoretical analyses in accordance with the theory. The whole previous history of science seems to indicate that these views will ultimately be absorbed into philosophical thought, no matter what the attitude of contemporary professional philosophers may be.

In this connection it is of particular interest to observe that the time contraction result (given in Equation 7 and used to explain the "slowness" of "moving clocks") has very recently received another striking confirmation in the observations of μ-mesons (heavy electrons) in cosmic ray showers. These μ-mesons travel on the average 10 kilometers from the place where they are produced in the atmosphere to the ground level where they are observed, whereas from their decay time (2 microseconds) and their speed ($0.995c$), they should be expected to travel only about 1 kilometer before decay. It turns out that the contraction factor $\sqrt{1 - v^2/c^2}$ in Equation 7 in this case is indeed just about one-tenth, which then provides agreement with the observed result.

This time contraction has a close connection with the clock paradox, which has also engaged the attention of philosophers ever since Einstein first enunciated it in 1911. The clock paradox is commonly presented in terms of the twins Peter and Paul. Peter at the age of twenty stays home on earth, minding his own business, while Paul decides to take a ride in a spaceship and visit a distant star. The ship cruises continually at a speed very close to that of light in free space. After about fifty years of this, as measured by Paul's faithful time-piece on earth, Paul returns to earth. Peter is then an old man of seventy, but Paul, because of the time contraction effect is still a vigorous youth, scarcely older than when he departed on his voyage. This result has been attacked as being repugnant to common sense, but as we have seen, this is scarcely a valid argument, since the meaning of common sense has altered much in the course of human history due to new discoveries in both physical science and philosophy.

A more logical assault on the conclusions about Peter and Paul has been based on the allegation that it is really contradictory to the very principle of relativity itself. This criticism has been put forth with great force by Herbert Dingle, who points out that the time contraction involved in Equation 7 is purely reciprocal, a fact that is readily substantiated and was well known to Einstein.[9] Thus if to A, in his inertial reference system, the clock attached to B's system (moving with respect to A with velocity v) appears to move slow, it also follows that to B in his inertial reference system the clock attached to A's system (also moving with respect to B with speed v, in the opposite direction) will appear to run slow. If it were not so, we could utilize this result to distinguish between the two inertial systems, and this would be contrary to the principle of relativity. To Dingle, the two twins in the paradox belong to such reciprocal inertial systems, and if either should age any faster than the other, it would be a violation of relativity itself. Opponents of Dingle's conclusions point out, however, that Peter and Paul are not strictly reciprocal systems in the ordinary sense of special relativity, for such systems pass each other once but never meet again.[10] In other words, whereas Paul and Peter have to suffer reciprocal accelerations in order to separate in the first place and join again after Paul's flight, Paul somewhere has to undergo an acceleration independent of Peter in order to make a cyclic trip and get back to where he started.

The matter of Peter and Paul's accelerations has caused some misgivings, since inertial systems do not undergo acceleration, and the deductions of special relativity apply strictly to inertial systems. According to G. J. Whitrow, the acceleration difficulty may be circumvented by assuming that the motion of Paul takes place in a finite universe of constant positive curvature (a Riemannian universe in the sense of our earlier discussion of space) and that Paul circumnavigates the universe before returning to Peter.[11] Whitrow can then show that Paul's clock will definitely run slower than Peter's in an absolute sense, since there is now an absolute difference between Peter and Paul in their respective relations with the universe as a whole. Some may object that this result depends on the assumption of a particular kind of universe and a particular kind of motion in it and that we do not know whether we live in this kind of universe; therefore, they contend, we cannot have complete confidence that it is a validation of Einstein's original presentation of the clock paradox. The matter continues to reside in the realm of controversy, though on the whole Whitrow and his adherents appear to have the better of the argument.

From the preceding discussion it is clearly important for physicists to examine and come to terms with the concepts of space and time, which they have commonly left to the scrutiny of professional philosophers. While Newton realized that the concepts of space and time were too important to be altogether overlooked by physicists, in general the language used by Kant and Leibniz to describe these concepts satisfied scientists for a long time. The value of scientific investigation of ideas normally deemed philosophical was finally shown by the work of Einstein and his contemporaries at the beginning of the twentieth century. Their work suggested very strongly that it is not practical to try to separate the concepts of space and time when using them in physics. They are concepts that enter science as inextricably mingled ideas, and the transformation equations from one reference system to another inevitably involve the symbols for both. That the principle of relativity is not a purely academic curiosity but has had a powerful role to play in the development of twentieth-century physics is shown by the fact that atomic physics could not have evolved without it. It is scarcely necessary to remind the reader that the famous Einstein mass-energy relation $E = mc^2$ was derived from the theory of relativity and is therefore connected with the physical attempt to come to grips with the philosophy of space and time. This should be enough to convince anyone that the philosophical aspects of physics are not without practical value.

Operationalism

We have just examined the way in which physicists look at the primitive ideas of space and time they employ in the construction of more elaborate concepts used in the development of physical theories. This leads to a more detailed analysis of how physical constructs come into existence than our discussion in Chapter Two, where the problem was touched on in connection with the epistemic and constitutive aspects of the definition of physical terms. One side of the problem has attracted much attention from both physicists and philosophers ever since P. W. Bridgman published his first book on the nature of physical theorizing.[12] Bridgman stressed with great force a point of view about the concepts of physics that has come to be called "operationalism." It is true that he often expressed his dislike for the terms operationalism and operationalist as implying an elaborate philosophical system, which he had no interest in setting up. He himself used the term operational analysis, but we shall employ the shorter term as sufficiently indicative of his point of view.

In its most restrictive sense the meaning of operationalism, as Bridgman described it, is that physical concepts or constructs shall be defined in terms of actual physical operations. According to this view a concept has no meaning unless it represents an operation that can be performed in the laboratory and so in this sense is instrumental. Thus it is meaningless to speak of the pressure of a gas until an operation is described that constitutes the actual measurement of pressure. The term itself then becomes a symbol to represent an operation or set of operations. For example, one might make a glass U-tube, fill it part way with mercury, leave one end open, and attach the other to a closed vessel filled with gas. A scale attached to the U-tube would then enable a number to be attached to the difference between the levels of the mercury in the two sides of the tube, and this number could be said to represent in appropriate, if arbitrary, units the gauge pressure of the gas (the amount by which the pressure of the gas inside the vessel is above or below the pressure of the atmosphere).

To continue with other examples we might say that temperature as a concept in physics is defined in terms of the set of operations by which one measures it by a thermometer in the laboratory. Mass is defined by the operation of using a balance, which enables us to assign a number to it. Electric current is defined by the operation of constructing an ammeter, which takes advantage of the heating or magnetic field associated with what we call the flow of current. We could enumerate many physical concepts that can be treated in this way. The point of view of the most thoroughgoing operationalist is that laboratory experiments must be involved in defining any concept used in physics. According to him it is definite and unambiguous. It tells us what a concept means in terms of a directive to go and do something specific: hence the appropriateness of the term construct. It is not merely verbal and therefore not susceptible of misunderstanding. It satisfies the best ideals of communication. A student can watch the operational procedure in any particular case and then proceed to carry it out himself. When presented in this light, the idea is a very attractive one, especially in a science that uses a host of concepts, which, if they were to mean different things to different people, could only lead to chaos.

However, another side of the picture becomes apparent once we have taken another look at thoroughgoing operationalism. Take, for instance, the definition of temperature. It is a matter of common observation that there are many kinds of thermometers—among them mercury-in-glass, alcohol-in-glass, constant-volume, and constant-pressure gas thermometers, as well as thermocouples, resistance thermometers, and radiation pyrometers. Which of these shall be used to provide a definition of temperature? Each thermometer involves the use of a highly specific set of laboratory operations. If we were to reply that it really makes no difference, since a measurement by any one of these thermometers under the same conditions will provide the same number, then the melancholy rejoinder has to be that this is unfortunately not the case in the world of our actual experience. In the same thermal environment a thermocouple and a liquid-in-glass thermometer—calibrated in the usual way on the centigrade scale at standard atmospheric pressure, by assigning 0 to the reading when the instrument is placed in melting ice and 100 to the reading in steam, and dividing the interval into a hundred equal parts—will not give identical readings in the same environment; the difference usually becomes greater as the temperature of the environment increases. Which thermometer then actually provides the correct temperature? Or must we replace the concept of temperature by a whole set of concepts depending on the instrument chosen and the set of operations followed? Even the rigid operationalist realizes that no coherent science can be devised by this procedure. The alternative open to him is to say that we must arbitrarily pick one set of operations to define temperature and then explain why the others do not agree. He is of course at liberty to do this, but unless he can present some reason for his choice that will be plausible to other physicists, his method is again an invitation to confusion and misunderstanding. He is unlikely to make a choice unaccompanied by some reason, but what sort of reason will he give? Already we are out of the realm of rigid operationalism. It seems almost certain that his reason will somehow be tied to physical theorizing, and indeed the choice of the constant-volume gas thermometer as a primary standard is closely connected with the thermodynamic and statistical-theoretical definitions of temperature.

We can make the case against rigid operationalism even stronger by noting that even experimental observations are never quite so definite that all experimenters will agree on how to employ a presumably definite and unambiguous recipe. At any rate there are many illustrations in the history of physics of misunderstandings over experimental processes. The operationalist is likely to reply that this is only adventitious, that as physics progresses experimental procedures become more readily described and prescribed.

The strongest case against rigid operationalism rests, however, on considerations not connected directly with experimental operations. If the only concepts allowed into physics were those defined in terms of actual laboratory operations, the construction of physical theories would be impossible. We have already commented on the difficulty encountered in defining temperature in terms of operations in the laboratory. The only reasonable way out of this is to construct a theory of thermal phenomena in which the concept of temperature is defined in terms of its relation to other elements of the theory. Thus in the statistical (molecular) theory temperature is defined in terms of the mean kinetic energy of the molecules. Of course it is necessary to identify this theoretical definition ultimately with temperature as measured by a thermometer. Both theoretical and operational aspects are necessary, as was shown in our discussion of the epistemic (operational) and constitutive (theoretical) aspects of the definition of a physical construct in Chapter Two. Without the constitutive aspects we have no basis for theoretical physics, and physics as a science falls to the ground.

The question now arises of how it is possible for any physicist to be a thoroughgoing operationalist in the sense we have attributed to the term. The obvious answer is that it is not possible, unless physics is to be reduced to pure empiricism and any attempt to explain phenomena with the help of theories is to be abandoned. Nevertheless, at one time it was thought that Bridgman really held this point of view.[13] Bridgman, however, declared that he had never intended to restrict operations to the purely physical, instrumental variety but that "mental" operations are not only allowable but necessary.[14] He went on to explain that by mental operations he meant "paper-and-pencil" operations. By this he evidently meant the constitutive definitions of theoretical physics—constructions of such concepts as the ψ or state function of quantum mechanics, the radius of an atom, or even the concept of energy itself in classical mechanics.

In 1950 Bridgman again laid stress on the importance of the idea of operationalism and applied his version of operational analysis to such concepts as field of force, action at a distance, and heat and entropy.[15] He again emphasized clearly the distinction between instrumental operations and paper-and-pencil operations and openly admitted the value of both kinds. It would appear that the "free construction" of concepts by the mind of the theorist is still permitted. Nevertheless, a careful reading of Bridgman inspires doubt about the actual latitude to be allowed to such free construction. A key passage in this connection is the following: "It will be seen that a very great latitude is allowed to the verbal and paper-and-pencil operations. I think, however, that physicists are agreed in imposing one restriction on the freedom of such operations, namely, that such operations must be capable of eventually, although perhaps indirectly, making connection with instrumental operations."[16] This passage seems to be evidence of a belief that ultimately the concepts of every successful physical theory must be instrumentally defined, unless the phrase "although perhaps indirectly" provides a loophole. If the phrase means that the concepts and postulates of a physical theory must always be of such a character that the laws deduced from them contain only quantities that are identifiable with laboratory operations, the problem disappears once more. But it is difficult to believe that Bridgman really meant this, for he would then be merely reiterating the conventional method of physical theorizing. No physicist is interested in producing theories predicting consequences that do not make definite connections with experience, and this means of course instrumental operations. In his 1950 lectures Bridgman seemed to be seeking valiantly to find some way of differentiating by instrumental means such concepts as, for example, action at a distance and action by a field. He concluded that there was no way to do this, the implication being that from the standpoint of operational analysis this was an unfortunate situation. (It must be admitted that he does not say this in so many words.)

It seems clear that a thoroughgoing operationalist will not be contented with paper-and-pencil operations that do not receive some instrumental backing. This harks back to the desire of the physicists of the seventeenth, eighteenth, and to a certain extent, nineteenth centuries to use in their theories concepts that are more or less directly connected with observable phenomena. (We have already commented on this in Chapter Two, in connection with the nature of physical hypotheses.) Thus, when Galileo wanted to study motion, he introduced concepts directly connected with the moving bodies he observed, such as velocity and rate of change of velocity. He also was much concerned with volume and density in dealing with large-scale objects. When he studied heat, he constructed a simple thermometer in which the temperature is measured by the height of a liquid column in a tube, the open, lower end of the tube being placed in the liquid container, its other end being closed by a bulb containing air. Their concern with material things colored the attitude of the seventeenth- and eighteenth-century scientists when they needed theoretical explanations of the phenomena of light, heat, electricity, and magnetism. For all of these phenomena they hypothesized theoretical media that had certain of the properties of observed material media but lacked certain others.

Seventeenth- and eighteenth-century theorists undoubtedly felt that they were sticking close to Nature herself in accordance with Francis Bacon's famous plea: "God forbid that we should offer the dreams of fancy for a model of the world!" To Bacon it was a scientist's duty to endeavor to understand nature in terms of the very stuff that experience presents to him:

"to dwell among things soberly, without abstracting or setting the understanding further from them than makes their images meet. . . . The capital precept for the whole undertaking is this, that the eye of the mind be never taken from things themselves, but receive their images truly as they are."[17] This has an impressive sound, as does much of the rest of Bacon's writings. It is not clear, however, that his recipe has been particularly successful in the actual development of physics, for even when the great physicists of the past thought they were trying to adhere firmly to concrete experience in framing concepts, they were exercising their imaginations vigorously. An example is Galileo's invention of the concept of instantaneous velocity. It was tied, to be sure, to ordinary experience, since it relates to the motion of particles we can see and handle. Nevertheless, it is an imaginative concept and required far more than ordinary observation to construct. It was by no means an obvious extension of ordinary observation, as is clear from the difficulty Galileo had in explaining what he meant by it to his contemporaries and convincing them that it was a valid and useful idea. Yet, of course, it is basic for kinematics and hence the whole of mechanics.

The history of physics is full of examples of this kind. One might suppose that the concept of energy, so fundamental for the whole of physics, came directly out of either observation or laboratory manipulation. But this was far from being the case. It took considerable intuitive vision to find an idea that would unify a whole aggregate of apparently dissimilar experimental observations, namely, the idea of invariance or constancy in the midst of change.

The thoroughgoing operationalist seems to have within him a kind of yearning for some ultimately valid and verified explanation of physical experience. It may even be that he is a strong believer in the reality of a world that is external to all observers and is an object of discovery for the physicist. In this case he might hold that theories are merely mental constructions for the purpose of assisting the process of discovery. According to this view, we can afford gradually to discard theories as we get closer to our understanding of the real world because we shall be dealing with things-in-themselves-and-as-they-are and not with things-as-they-might-be. A further tenet of this view is that true scientific understanding will ultimately have been reached when all phenomena form an interlocking set of relations that are purely instrumental in character; the task of science will then have been completed.

It is not indeed at all likely that Bridgman held this point of view, nor does any modern physicist of repute appear to. But probably many earlier physicists and philosophers had a leaning toward it, and our descendants may return to it, particularly if one set of theories should prove so successful as to inspire confidence that all future experience could be fitted to it. There are persons who believe that our present atomic theory is on the right track to such an extent that necessary changes in the future will be minimal. It is difficult to understand this feeling of assurance, for it seems to ignore the lessons of the history of physics as well as the frightful speed with which new physical experience is being created on a grand scale.

Thoroughgoing or rigid operationalism appears to be outside the mainstream of successful physical inquiry. The tempered operationalism of Bridgman's middle years, which incorporated mental or paper-and-pencil concepts along with those instrumentally defined, seems to be nothing more than the standard scheme of physical theorizing now current, with varying degrees of emphasis, among modern physicists. The merit of the operational point of view is its emphasis on the importance of the process of identification of the results of physical theorizing with actual experience. This means that among the concepts of physics there must always be some that have both epistemic (operational) and constitutive (theoretical) aspects in their definitions. Otherwise the theories of physics would fail to make contact with experience. For this contribution of the operational point of view we ought to be grateful.

Causality and Determinism

Causality has been the subject of philosophical inquiry for ages, usually in terms of the apposition of cause and effect. That nothing ever takes place without a cause seemed evident to the earliest thinkers. A cause was defined as an occurrence that precedes the given effect, that is always associated with the appearance of the effect, and that is never absent when the effect appears. But many difficulties soon appeared in this simple-minded point of view, among them that in connection with a given phenomenon a number of occurrences often satisfied the conditions set forth, and the question arose as to which condition should be called the real cause. From the very nature of physics apparent causal relations among physical phenomena are continually encountered, and so it was inevitable that scientists in general and physicists in particular would have to take some sort of attitude toward this philosophical problem. An illustration of apparent cause and effect in physics is the decrease in the volume of an ideal gas at constant temperature when the pressure is increased. It is tempting to say that the increase in pressure is the cause of the decrease in volume. However, simple reflection shows that it is just as sensible to say that what really happens when the gas is squeezed is that the volume is decreased—the decrease then becoming the cause of the observed increase in pressure. The notion of cause and effect in this case thus appears to be of little value. Yet the unfailing existence of the relation between pressure and volume does suggest a situation analogous to the philosophical notion of cause and requires the introduction of a principle of causality.

This principle is simply an expression of confidence that physical experience is not a wholly chaotic affair but that order prevails and can be represented by relations connecting elements of experience—physical laws. The idea of causality further assumes that the form of such laws is not an explicit function of the time. In other words, if we repeat an experiment on the equation of state of a given gas next week or next year, we expect to find the same result as that obtained today (barring, of course, minor variations attributed to experimental error). Moreover, we also expect that the form of the law should be independent of the place where the experiment is performed. "Constants" of nature may change in space and time, but this should not affect the form of physical laws: the acceleration of gravity will change from place to place and may even change with the time at a given place, but this should not change the form of the law of falling bodies. The principle of causality in physics is simply a reflection of our confidence that there really is regularity in physical experience and that we are not deluding ourselves when we try to set up physical laws.

Closely related to the concept of causality is that of determinism, and in the past there has been a tendency to equate the two ideas. Since a given specific cause must of necessity result in a specific effect, the effect is determined; however, as far as physics is concerned there is an important difference between causality and determinism.

A physical theory is said to be deterministic in character if it makes possible the prediction of the future course of a given phenomenon. The simplest illustration of determinism is provided by the theory of classical mechanics. If we know the state of a mechanical system at any one instant, mechanics enables us to predict all past and future states of the system; we may say that the whole temporal behavior of the system is completely determined. The solar system, consisting of the sun and planets and their satellites, constitutes such a system to a good approximation. Knowledge of the state of the system at one instant—of the position and velocity of every body in it (strictly, in practice, of the positions at two successive instants)—permits the calculation of the position of every body in it for all future times and indeed for all past times also. Hence tables of planetary positions can be prepared with assurance of precise prediction. The degree of precision is evident in the accuracy with which astronomers can pinpoint the time of a total solar eclipse, which is caused by the relative positions of the sun and the moon as seen from the earth. It is customary to fix in advance the onset of totality within a few seconds of the actually observed time.

The success of determinism in celestial mechanics, in the related field of terrestrial ballistics, and more recently in missile and artificial satellite technology has made a powerful impression on all who use mechanical concepts and can understand how they work. Certainly the thinkers of the eighteenth and nineteenth centuries, impressed by the success of Newton and his followers, tended increasingly to identify determinism with causality, thus introducing a certain amount of confusion into the methodology of modern physics.

The advent of the kinetic theory of gases, postulating the existence of molecules as the constituent entities of a gas, obviously produced difficulties for the deterministic point of view. In order to account for the observed properties of the gas, it is necessary to assume that the number of molecules per unit volume under standard conditions is very great indeed (of the order of 10^{19} per cubic centimeter). Even if it is postulated that these molecules move in accordance with the principles of classical mechanics, it is clear that, owing to their large number, the practical assignment of the initial conditions (the state of the whole system) at any instant for all the molecules is quite out of the question. Even if this were possible, the application of the boundary conditions at the walls would introduce another obstacle to the fixation of the state. Hence for such a system determinism, while possible in the ideal sense, is illusory. We must then fall back on statistical considerations and estimate the average behavior of the collection of molecules.

By calculating the average rate of change of momentum undergone by the molecules in a unit volume by collision at the walls and expressing the calculations in terms of the average velocity of the molecules, it is not difficult (if suitable approximations are made) to derive relations connecting average quantities. (It has to be admitted that there are various equally reasonable ways of taking averages and the choice is arbitrary; within rather wide limits, however, it turns out that the choice of average is not critical.) The remarkable result is that—by suitable and plausible identification of the average quantities with experimentally measurable ones—the deduced relations become recognized laws of gas behavior, like the equation of state of an ideal gas. Thus, by abandoning determinism as far as the individual molecular behavior is concerned and by concentrating on the average behavior of the whole aggregate, we are able to derive laws descriptive of the physical phenomena exhibited by the gas represented by the aggregate. In the light of this result we are entitled to call the theory causal in character, though it is not deterministic in the sense that the theory of classical mechanics is. This example shows the possibility of the existence of causality without determinism and suggests that the two ideas are not synonymous.

The use of statistical reasoning is fraught with peculiar difficulties, for it inevitably involves the introduction of the notion of probability. Probability, one of the trickiest ideas that human minds have ever played with, is subject to almost continual controversy among mathematicians and philosophers.

Any statistical average is of course subject to fluctuation: the average number of heads that appear when an ordinary coin is tossed a great many times is expected to be equal to the average number of tails; but there is nothing to prevent a long run of heads and no tails. (This, to be sure, is not expected often, and mathematical computation reinforces common-sense expectation.) The larger the number of cases over which a statistical average is calculated, the smaller is the average fluctuation from average behavior. Since the number of entities (molecules, for instance) entering into statistical theories in physics is usually enormously greater than unity, there is some basis for confidence in the validity of the statistical method. At any rate it is a method that makes it possible to understand many of the observed properties of matter in a way that no other point of view comes anywhere near satisfying.

With statistical reasoning we can calculate the specific heats of gases and the effect of the temperature on the viscosity and predict the viscosity's independence of the pressure. We can also establish a relation connecting the viscosity of a gas with its thermal conductivity. Applied to the electrons in a metal, statistics permits the deduction of the Wiedemann-Franz law connecting the thermal and electrical conductivities. Physicists have therefore frequently resorted to statistical reasoning in spite of its indeterministic character and will doubtless continue to do so in the future whenever systems of large numbers of entities are involved. Since it always leads to relations that can be identified with experimental laws, we feel fully justified in referring to it as being of causal character.

The situation with respect to determinism is quite different in the case of the behavior of individual atoms as described in the quantum theory. In 1900 Max Planck finally obtained a theoretical explanation of the experimental law describing the distribution of energy in the thermal radiation spectrum of a black body. He was able to do this by assuming that the energy values of the radiators in equilibrium with the radiation field form a discrete rather than a continuous sequence. This was later interpreted to mean that energy is emitted by a radiator in lumps, or quanta, depending on the frequency f, and indeed the quantum of radiant energy of frequency f is taken in Planck's theory to be equal to hf, where Planck's constant h is a fundamental constant of nature (in the cgs system of units, 6.57×10^{-27} erg sec). The discontinuity in the emission of radiation involved in the Planck theory is very small on the macroscopic level, but it becomes important when single atoms are in question. The theory rapidly took hold in the developing atomic physics of the early part of the twentieth century, probably attaining its highest success in Niels Bohr's theory of atomic structure. Bohr was able to predict many observed regularities in atomic spectra and other atomic properties with remarkable precision. Although Bohr's laws involve Planck's constant h, they are just as causal in character as the observed regularities of classical mechanics and satisfy all the requirements of causality discussed in the earlier part of this section.

In Bohr's early representation of his theory in terms of the Rutherford nuclear atom model, he used a peculiar mixture of classical mechanics and nonmechanical ideas. Only certain states of motion of the electrons about the nucleus were permitted (the so-called stationary states), whereas classical mechanics imposed no such restriction. Moreover, radiation from the Bohr atom results only when an electron makes a transition from one stationary state to another and loses energy. The frequency of the emitted radiation is equal to the energy loss divided by Planck's constant h. As far as an individual atom is concerned, the theory is unable to predict when a transition will take place. It can only estimate the average time between transitions (of the order of 10^{-8} sec), and on this level indeterminism seems to be integral to Bohr's theory. But it is not like the indeterminism of the molecular theory of gases, where a theoretical determinism is possible. Although the behavior of the single atom with respect to energy change is a matter of probability estimate only, this does not adversely affect the power of the theory to produce causal laws that exhibit regularity in experimental phenomena. The indeterminism in the finer details of the Bohr theory might not therefore have been expected to cause much of a stir. The contrary, however, has been the case, and the ensuing situation is an interesting episode in the history of physics.

When certain shortcomings were detected in the Bohr model in the early 1920's (mainly in connection with the spectra and other properties of polyelectronic atoms—atoms that in the normal condition have more than a single electron revolving around the nucleus), two new views were introduced. One was the wave mechanics of Louis Victor de Broglie and Erwin Schrödinger; the other was the matrix mechanics of Werner Heisenberg. The former has a pictorial quality in that the behavior of an atom is described in terms of waves, which, though not in themselves electromagnetic waves, yet are analogous to other waves in physics. They satisfy a second-order partial differential equation called the wave equation with solutions satisfying certain boundary conditions, just as transverse waves in a string satisfy the end conditions and lead to normal modes or characteristic frequencies. Both authors of this point of view probably believed that its use would make determinism possible on the atomic level in the quantum theory of atomic structure.

Heisenberg, on the other hand, decided that the effective way to develop the quantum theory—or quantum mechanics, as it came to be called—was to forgo any attempt to work with a picture but merely to represent states and behavior of atoms in terms of purely mathematical expressions. Since the quantities to be worked with were best represented by matrices, Heisenberg's version of quantum mechanics became known as matrix mechanics. It derived the energies of the stationary states of atoms and was causal in character. Nevertheless, in order to estimate the probability of transitions between such states (leading to emission or absorption of radiation), it was necessary to use probability considerations. Hence it reflected the indeterminism of the earlier Bohr model, though it was successful in tackling problems that the Bohr scheme could not solve.

As far as space and time are concerned, there should be no problem about the behavior of individual electrons or other elementary particles, because these considerations do not enter into the matrix theory. Heisenberg's original point of view made no provision for a picture of an electron —or a whole atom, for that matter—doing anything in space and time; there was strictly no meaning to the position of an electron at an instant or to its change of position with time. In a mechanical sense the indeterminism was of no consequence. However, Heisenberg was ultimately impelled to apply his theory to the behavior of the single atomic particle in order to satisfy the requirement of picturesqueness and so came up with the celebrated indeterminacy principle. It is often asserted in popular books that indeterminacy means the fundamental inability to measure simultaneously the position and momentum (mass times velocity) of an atomic particle. It is said that if light of very short wave length is shone on the particle to determine its position, the effect of this radiation is to give the particle an unpredictable amount of increase in its momentum. This picturesque way of putting it has an element of validity, but it overstates the actual deduction from the theory of quantum mechanics, which states that the product of the root-mean-square deviation from the mean of a set of position measurements and the similar quantity for a set of momentum measurements is equal to, or greater than, Planck's constant h divided by 4π. Each root-mean-square deviation (or standard deviation, as it is usually referred to in statistical studies) is called the indeterminacy in the quantity in question, whence the name of the principle. It is clear from the formulation that if the indeterminacy in position is represented by δx, the indeterminacy δp in momentum is at least $h/4\pi\ \delta x$. If the indeterminacy in position could by some increase in precision of measurement be reduced to zero, the indeterminacy in momentum would become infinite, and the momentum could not then be measured with any precision at all. It is possible to deduce this principle from the postulates of quantum mechanics.[18] But for our purpose it is sufficient to note the conclusion that it is impossible to measure simultaneously both the position and momentum of an atomic particle (strictly speaking, any particle to which quantum mechanics applies) with infinite precision (zero indeterminacy). This conclusion is the basis for the statement that the behavior of the atomic particle is indeterminate and hence that atomic physics is basically indeterministic.

In as far as quantum mechanics applies to particles of all kinds, this indeterminacy strictly applies to all particles encountered in experience.

However, this poses no real difficulty for determinism in macroscopic mechanics, since for such particles of comparatively large mass (compared with atomic particles) the amount of indeterminacy is very small compared with that already involved in the normal errors of measurement. Such errors do not interfere with the deterministic character of classical mechanics: we always have to develop an arbitrary theory of errors and estimate the appropriate precision of measurement in the case of any macroscopically measured quantity before using it in any theoretical calculation. It is of course possible to calculate the indeterminacy in position and momentum of any atomic particle by the methods laid down in the theory of quantum mechanics. This might seem to be adequate for all practical purposes, but the precision of the calculation is affected by the magnitude of Planck's constant h, which assures that in the case of a particle like an electron the indeterminacy may well be of the same order as the magnitude of the quantity being measured. Another difference between the classical-macroscopic and atomic cases is found in the fact that though in the classical case theoretical ways can be devised to compensate for the inevitable errors of measurement, there seems to be no way to do this in an unambiguous fashion for atomic particles. And so we are forced to admit fundamental indeterminism into atomic physics if we accept the theory of quantum mechanics. This theory has been so successful in handling atomic structure problems that its abandonment is unlikely.

Nevertheless the urge toward determinism in the construction of physical theories is great, especially when the entities in question are single particles. Unfortunately Schrödinger's attempt to maintain determinism in the behavior of single particles by associating a wave with the particle failed. In more recent times the young American theoretician David Bohm has made a somewhat different attack on the problem. According to him the Copenhagen school of quantum mechanicians inspired by Bohr, which insists that indeterminism associated with the indeterminacy principle is intrinsic in nature and can in no way be avoided, is proceeding on the unwarranted assumption that the present form of the quantum theory must continue to be accepted without change.

The indeterminacy principle is a direct deduction from the fundamental postulates of quantum mechanics and owes its logical status wholly to quantum theory. Indeed, it would have no basis for existence outside the theory, and the interpretation of the principle in terms of the disturbance produced in an atomic system by the attempt to conduct a measurement has no logical quantum status. Such disturbances occur in all physical measurements, and there are classical methods of compensating for them. The basic significance of h in this connection enters only if the quantum theory is postulated. John Von Neumann and others have attempted to show that to restore determinism to atomic physics is logically inconsistent with the indeterminacy principle; however, as Bohm has indicated, if a new theory is developed to replace quantum mechanics, the whole theory of determinism in atomic physics will have to be restudied.

Bohm and others have advocated the invention of a new theory, and its technical details can be found in the papers of Bohm and of his co-workers and sympathizers.[19] The general idea of the new theory can be understood in terms of the Brownian motion, in which small particles of smoke suspended in a gas under equilibrium conditions are observed to perform random displacements; these displacements are interpreted as due to the impact of the molecules of the gas. The molecular theory of the Brownian motion predicts that the root-mean-square displacement of a given particle varies directly as the square root of the time of observation, while the root-mean-square fluctuation in momentum varies inversely as the square root of this time. Hence the product of what may be called the average indeterminacy in position and the average indeterminacy in momentum of the Brownian particle is a constant independent of the time. This forms a kind of indeterminacy principle analogous to that of quantum mechanics. There is therefore an indeterminacy in the Brownian motion, and this is intrinsic at the level of observation of the motion—nothing can be done at that level to remove it or compensate for it. However, at the lower level of the molecular motion responsible for the motion of the Brownian particle the indeterminacy can theoretically be removed: the motions of the individual molecules are theoretically determined by the principles of mechanics and the boundary conditions imposed by the container of gas. In statistical mechanics, of course, this determinism is forgone as a matter of convenience, and statistical averages are used.

In the theory of classical mechanics there is nothing intrinsically indeterministic about the behavior of any particle in an aggregate, no matter how large the aggregate may be; in quantum mechanics, on the other hand, the behavior of every atomic particle is intrinsically indeterministic. According to Bohm, just as the indeterminacy of the Brownian motion on its level of observation may be explained on the sublevel of an intrinsically determined molecular motion, so the indeterminacy of quantum mechanics may be explained on the sublevel of a new theory in which the entities are as "hidden" from observation as the molecules of a gas are hidden from direct observation. Moreover, the entities or variables in this new theory need not be completely undetermined. We could at any rate have freedom of choice and might, if we wished, remove the intrinsic fundamental indeterminism associated with quantum mechanics as presently understood by the Copenhagen school. Although others have seized on Bohm's theory as an out-and-out attempt to restore determinism to quantum physics, in his own view this question remains open and is subsidiary to the possible advantage of his theory in ultimately providing a more successful explanation of nuclear behavior than is afforded by present-day quantum mechanics.

It seems only reasonable to keep an open mind in judging the question of determinism versus indeterminism in quantum physics. The strong adherents of the Copenhagen school have clearly adopted too dogmatic an attitude, and such an attitude in general is apt to be unfavorable to the successful development of physics.

Probability and Statistics

In the previous section we commented on the fact that the introduction of statistical averages in the molecular theory of gases necessitates use of the concept of probability in physics. In some ways this seems to be a pity, for probability has given great difficulty both as a primitive, intuitive notion and as a more precisely defined construct. However, physicists cannot avoid it.

The fundamental question involves the meaning of the word probable, a meaning that has been examined over the years by many celebrated philosophers and scientists. (Pascal, Leibniz, and Laplace were among the more important in relatively modern times.) Of course we all use the concept every day in a vague sort of way, as when we say: "It will probably be fair tomorrow," or "I shall probably get a haircut today." Both cases indicate a wish to predict an occurrence of interest to us but express an uncertainty about it based on ignorance of all the factors involved. Uncertainty and ignorance are definitely involved in the concept of probability. In everyday life the indefiniteness of the concept scarcely bothers us, but trouble begins when we ask that a numerical value be given to probability. Those whose business it is to make money by predicting uncertain events wish to know how to reckon their chances. In fact, the modern mathematical theory of probability appears to owe its origin to questions about games of chance proposed to Pascal. Insurance companies must have some means of calculating the premiums their policyholders should pay. The manufacturer of an item to be sold in large quantities wants to know how large a sample he must test for defects in order to insure that the number of defects in his whole output shall be kept below an assigned number. These are some of the many practical uses for a numerical theory of probability. The physicist also needs such a numerical theory so that his measurements will have a meaning.

Many attempts have been made to define probability in mathematical terms, but here we shall discuss only some aspects of probability that relate particularly to the needs of physics and the development of physical theories. In doing so we can safely take refuge in the conventionalism of Poincaré, for the precise mathematical definition of probability fortunately appears to have little consequence in physics. In the light of what has been said about the importance of the concept, however, this statement requires clarification. The principal use of probability in physical problems is to calculate averages of quantities representing properties of aggregates of large numbers of entities. These are called statistical averages. Suppose, for example, that the problem is to evaluate the average energy of an aggregate of a large number of particles, as in the molecular theory of gases. Such an average can have several meanings. We are all familiar with the arithmetical average of a set of numbers—the sum of the numbers divided by the number of the set. The utility of such an average is that it provides a single number that can somehow represent the whole set of individual numbers for certain practical purposes. If the numbers in question are the values of a certain physical quantity obtained by successive measurements, the problem is to find a single value to represent the result of the measurement—an arbitrary procedure but a convenient one. In many cases the arithmetical average is satisfactory; however, if we have reason to believe that not all the measurements have the same validity, we may not wish to assign the same weight to each one. Hence in calculating the average in question, we must multiply each measured value x_i by a real number P_i (the probability coefficient that is associated with it) and total all the products. The average value of x, or the value we shall accept as the result of the measurement, is then defined as:

$$\bar{x} = \sum_{i=1}^{N} P_i x_i , \qquad \text{(Equation 8)}$$

where N is the number of measurements. It is to be understood that Equation 8 holds only if the probability coefficients P_i are positive real numbers of such a character that

$$\Sigma\, P_i = 1. \qquad \text{(Equation 9)}$$

They are therefore all proper fractions or in no case greater than one.

The main problem in the applications of probability in physics is to choose the probability or weight coefficients P_i. From the qualitative meaning of the concept, it is clear that if a particular value x_i is more likely to occur than another value x_j, then P_i is greater than P_j. This is not a very specific requirement, and it is clear that a great many functions that satisfy it can be concocted. In one very common case large values of x are less likely than small values of x, as, for example, when x is the deviation from the expected value in a set of measurements. In this case a favorite probability function is the Gaussian, or "normal error," function,

$$P_x = \frac{2}{\sqrt{2\pi n}}\, e^{-2x^2/n}, \qquad \text{(Equation 10)}$$

where n is some constant giving the range of x. (Here x is confined between $-n/2$ and $+n/2$.) If n is taken to be sufficiently large, it may be assumed to be effectively equal to infinity as far as summation is concerned, and then the formula

$$\int_{-\infty}^{+\infty} P_x\, dx = 1 \qquad \text{(Equation 11)}$$

is in agreement with Equation 9.

Much argument has been devoted to the project of demonstrating the inherent plausibility of the Gaussian probability function, or distribution law, as it is often called. This could be avoided by recognizing frankly the arbitrary character of the law and treating its use as a matter of convention as far as many physical applications are concerned. In this respect the distribution law, or function, might well be considered in the same logical light as the force function in classical mechanics. We use different functions to describe different observed types of motion, and the choice is a matter of judicious guesswork in which, of course, we are guided by previous experience.

When probability distributions are used in statistical mechanics, as in the Gibbs ensemble method, the choice of ensemble probabilities need not be decisive in the calculation of averages. Gibbs used both the microcanonical and canonical ensembles successfully in the calculation of averages: in the former case the ensemble consists of a part of phase space in which the total energy of the system being studied has almost the same value throughout; in the latter case a much larger part of phase space is used, and the energy varies widely. It turns out that the choice of ensembles and corresponding probability distributions does not in this example make much difference to the eventual averages of quantities in the description of the behavior of the system. The important point is that the use of probability distributions in statistical mechanics is essentially arbitrary and conventional and that elaborate theory as to the meaning of probability in this discipline is therefore unnecessary. It may of course be argued that statistical mechanics is not the only part of physics that employs probability and moreover that a failure to come to grips with a thoroughly logical theory of probability may ultimately lead to inconsistencies in its employment in physics and hence to frustration.

There have as a matter of fact been numerous criticisms of statistical mechanics based on the conviction that no matter what probability distribution is used, the calculation of averages may be rendered meaningless by the inevitable fluctuations in the quantity whose average is sought. When a given value of the quantity is assigned to a certain probability, the value may be realized five times out of a hundred; but from the very nature of probability, it may on occasion be realized six or even ten times out of a hundred. There are bound to be fluctuations that will bias the average. One answer to this criticism is that an average deviation from the average can always be calculated to provide a kind of measure of the validity of the average itself. This important contribution of Gibbs's formulation of statistical mechanics provided assurance that the averages calculated from it were physically meaningful, for he found that in the case of a canonical ensemble the average deviation from the mean values of physical quantities characteristic of the system becomes negligible as the number of degrees of freedom of the system grows sufficiently large. To the mathematician or philosopher who worries over the fundamental meaning of probability and the validity of its use in physical problems, even this result will not be too reassuring; but it has satisfied most physicists.

Measurement

In physical experimentation the idea of measurement involves the introduction of the further construct of a physical quantity, which in turn requires symbolism for its expression. Thus a set of operations intended to measure the quantity symbolically called pressure results in attaching a number to the quantity. This numerical symbolism not only has the merit of economy in describing portions of experience, but it also calls into play all the resources of mathematics for the description of nature. Though measurement has often incurred the scorn of humanists, who insist that it distorts experience, it is one of the most powerful tools ever invented in the human attempt to enlarge experience and make it more meaningful.

However, the concept of measurement has raised problems that have become particularly important in contemporary physics. The performance of measurements inevitably led to the development of the idea of a measuring instrument—a physical object or collection thereof whose purpose it is to measure a physical quantity characteristic of some other object. In the measurement of pressure, for example, it was tempting to dissociate the U-tube partly full of mercury from the vessel to which it was attached and to think of the tube as a pressure gauge for measuring the pressure of the fluid in the vessel. There is no logical basis for this separation, for the whole set of operations involving the construction of the tube and its attachment to the vessel, together with the reading of the difference in the levels of the mercury in the two sides, constitutes the measurement. However, the dissociation became a matter of some convenience as soon as it was found that the same physical device, attached to different vessels containing different fluids, could be used over and over again with consistent results. The validity of this procedure was strengthened when the development of theory showed that arrangements of quite different physical appearance and behavior could also measure pressure, if suitably attached to the appropriate vessel.

Subsequent experiments strengthened the view that every physical measurement is performed by means of an instrument consisting of a set of physical objects specifically designed to measure some property or state of the environment in which it is placed. This gave rise to the concept of a pressure gauge for the measurement of pressure, a thermometer for the measurement of temperature, and all the various types of electrical and mechanical meters that have played such an important role in classical physics. The development of measuring instruments emphasized the distinction between the object being observed and the instrument doing the measuring—the instrument being considered as a kind of tool for extending and refining the senses of the observer. This distinction strongly suggests the division of the domain of experience into a world of physical objects on the one hand and a collection of individual observers on the other—in accordance with the common sense view of a real and objective universe, which it is the task of science to describe and understand. It is of course not so readily harmonized with the view that experience is a whole, that it is constantly being increased, and that any attempt to divide it into segments like object and observer is artificial and may well lead to unfortunate consequences as new experience is created.

Nevertheless, the idea of a measuring instrument has worked well throughout the history of classical physics. It was early recognized, to be sure, that the act of attaching an instrument to an object or immersing it in a given environment always changes the object or environment from its state before the instrument was put in place—hence it is logically impossible to obtain a measurement of an isolated object. However, this has posed no insuperable objection to the object-instrument view of measurement, since in many cases the influence of the instrument on the object can be shown by theory to be negligible compared with the magnitude of the measured effect. And in all cases, even if the effect is not negligible, it can be calculated and allowed for in the final assessment. Thus when a meter stick is used as an instrument for the measurement of the length of a table, it is hard to believe that merely placing the stick on the table has any appreciable influence on the length, though of course a highly elaborate theoretical study in which gravitational forces are considered would show the existence of minute reaction effects. On the other hand, placing an electrical meter in a circuit always changes the resistance of the circuit by a readily calculable amount.

Further complications arise with the object-instrument concept in the attempt to measure constructs of a purely theoretical or constitutive character. Such a construct might be the radius of an electron orbit in the Bohr model of the hydrogen atom or the dimensions of the electron itself in the various types of theories of this fundamental concept. Even simpler would be the concept of the position or the velocity of an electron either in an atom or free in space. Since instrumental methods exist for making such measurements on the macroscopic level, it is tempting to try to invent mental operations, which, if they could be carried out, would yield measured values on the microscopic level. The attempt to do this has had interesting consequences. In the first place, this procedure is unnecessary for the development and testing of the related theory. To be satisfactory the theory itself must provide relations between such constitutive constructs and other constructs that can be epistemically defined. In this way perfectly valid numerical values can be assigned to the constitutive quantities, even if they are not directly measurable. Thus the radius of a Bohr orbit is related theoretically to the energy states of the atom, and these states in turn are connected with the possible frequencies of emitted and absorbed spectral lines. Since spectral lines are measurable by instruments called spectrometers, values of the radius are readily calculable, and from these values theoretical estimates of the dimensions of an atom can be obtained. This seems straightforward enough and in accord with a reasonable view of operationalism as set forth earlier in this chapter. It does not outlaw the theoretical use of purely constitutive constructs, so long as they are theoretically related with constructs corresponding to epistemically defined quantities.

Yet it must be admitted that the persistent assignment of instrumental significance to purely mental measurements has had a considerable effect on the logical interpretation of modern physical theories of atomic structure. Suppose, for example, the problem is to measure the position of an electron by a mental operation. In order to perform this operation for a large-scale object it is necessary to see the object, which means that light must be directed on it and scattered or reflected in such a way as to make the object's position relative to a scale unmistakably clear. With an object of what may be called normal dimensions (of the order of magnitude of our own bodies) visible light is adequate for satisfactory seeing. The theory of optics suggests that the ability of electromagnetic radiation to locate the position of an object is given by the relation between the size of the object and the wave length of the radiation. If the ratio of wave length to size is large compared with unity, the object will be only diffusely defined, whereas if the ratio is much less than one, the location will be sharp. This conclusion is not only verified in the case of ordinary light and normal objects but also by the successful use of X rays in the study of the distribution of matter in the form of "atoms" in a crystal.

In contemplating the measurement of the position of an electron, it is therefore natural to perform the theoretical operation of shining light of ultrashort wave length (gamma rays) on the electron. Ordinary gamma rays actually have a wave length still too long compared with the presumptive theoretically deduced dimensions of the electron. However, in modern atomic accelerators (atom smashers) it is possible to give elementary particles energies up to more than a billion (10^9) electron-volts. The electromagnetic radiation corresponding to this would have a wave length of about 10^{-13} centimeter, and this would be adequate to pinpoint an electron. But now we have to reckon much more seriously with the interference of the measuring instrument (the radiation) with the measured object (the electron). In the act of shining on the electron, a single photon of the light in question will communicate enough energy to the electron to move it suddenly from its original position, and the electron will be unable to stay in a condition of equilibrium long enough to be measured. The concept of equilibrium in the measuring process is closely tied up with the extent of the influence the measuring instrument has on the measured object. Unless a system is in equilibrium while the measuring process is going on, it is difficult to attach meaning to the result without an elaborate process of allowance-making.

It is not difficult to see the possible relation between the preceding mental experiment and the celebrated indeterminacy principle in quantum mechanics. But the principle of indeterminacy, which derives primarily from the characteristic definition of state in quantum mechanics, ignores the influence of the measuring instrument on the thing measured. Nevertheless, the theory involves the somewhat arbitrary definition of a mental measurement of what are called quantum mechanical observables, and great emphasis is placed on the relation between the average deviation from the mean in the measured value of one observable and the similar value for the canonically conjugate variable (momentum and position, for example). This elaborate theory of measurement in quantum mechanics has resulted specifically from the object-instrument point of view of measurement in classical physics. In fact, Niels Bohr essentially made the object-instrument idea the basis of his famous complementarity principle. According to Bohr, because of the division of all human experience into objects to be observed and observers to make the observations, all human description must be in terms of mutually exclusive modes, which indeed exhaust the possibilities of description. Consider the case of the dynamics of the electron. In classical physics the dynamic state of the electron is completely determined by its position and velocity at any instant, and it is assumed that both these quantities can be measured instantaneously with arbitrarily great precision. In the quantum theory, on the other hand, this is no longer the case, and the complementarity principle implies that position and velocity (strictly, momentum) are mutually exclusive observables. It is impossible to have simultaneous knowledge of the position and of the velocity, although knowledge of the two is exhaustive as far as description is concerned. They are mutually exclusive aspects of our experience of electrons. Bohr was so impressed by what he thought was the ineluctable consequence of the object-observer interpretation of experience that he discovered numerous other illustrations of the complementarity idea in other domains of experience. Thus in biology the description of the living organism is exhausted in terms of the mutually exclusive ideas of analysis on the one hand and over-all behavioral observation on the other. In analysis the biologist tries to understand the nature of life by dissecting the organism into its constituent parts and endeavoring to understand the behavior of the whole organism in terms of the functions of these constituents. In the latter case he takes the holistic point of view and examines all aspects of the behavior of the organism as actually observed. The mutual exclusiveness of the two modes of description is exemplified by the fact that any attempt at a complete analysis leads inevitably to a dead organism, whereas the holistic scheme forgoes knowledge of the role of the component parts.

The idea of complementarity is intriguing, and it is easy to understand how a profoundly thoughtful philosophical physicist like Bohr should have been so impressed by it. He even extended the complementarity idea to the field of social phenomena and, for example, looked upon the concepts of love and justice as complementary. Yet a candid appraisal shows that complementarity is entirely dependent on the assumption of a necessary object-observer dichotomy of description in physics. The object-observer principle rests on the concept of the measuring instrument, and if measurement is conceived rather in terms of actual operations, this dichotomy can be avoided. It may be objected that we cannot really get away from the object-observer point of view in our theories, because theories are based on concepts related to objects. But this is not strictly true, for the constructs and postulates of a theory can be as abstract as we like, as long as we ultimately provide for an identification between the results of the theory and the operational procedures involved in testing it. As a matter of fact, quantum mechanics was devised as an abstract theory of precisely this character, and as we have seen, it was only the desire to introduce a pictorial quality into it that led to indeterminacy and complementarity. It is doubtful whether the price that physics has had to pay for this is justified by any possible advantage in the results. We should avoid dogmatism in these matters.

Since all measurements involve errors, the development of a theory for handling such errors has been necessary; this involves probability. Repeated measurements of the same quantity under what are supposed to be the same conditions always yield numbers that differ to a greater or lesser extent from each other. The practical problem thereupon arises as to what number should be assigned to the quantity being measured. If the scatter is not very great—if the greatest difference between any of the measured values is very much less than their arithmetical mean—it has in general been thought to be satisfactory to take this mean value as the result of the measurement. But questions immediately arise. Why use the arithmetical average, when there are many other types of averages—for example, weighted ones based on the idea that large deviations from any hypothetical value for the measurement are less likely than small deviations? Thus in a series of measured values the assumption can justifiably be made that the extreme values are less likely than those in the middle of the series. (Merely adding all the numbers and dividing by their number might weight unduly the extreme values.)

The use of the word likely in the preceding paragraph indicates that we are dealing with probability. Since there are many ways of defining probability, we might conclude that there is no categorical method for assigning the result of a given measurement. However, things are not quite so bad as all that, and also we have to have a definite experimental result. It turns out to be appropriate, excluding the possibility of systematic errors due to defects in equipment, to take as the probability of finding a value x for a given quantity subject to random errors the expression $P(x-a)dx$, where $P(x-a)$ is the Gaussian distribution function given in Equation 10 with x replaced by $x-a$. We wish a to be the hypothetical "correct" value of x and hence evaluate it as that value which maximizes the expression

$$P(a, x_1 \cdots x_n) = Ce^{-b\left[(x_1 - a)^2 + (x_2 - a)^2 + \cdots + (x_n - a)^2\right]},$$

(Equation 12)

where $x_1 \cdots x_n$ are the n measured values in question. Analysis then shows that the value of a for which P in Equation 12 is a maximum is

$$a = \bar{x} = \frac{\Sigma \, x_i}{n} , \qquad \text{(Equation 13)}$$

or the arithmetical mean of the separate measurements. This means that if we assume a Gaussian probability function for the distribution of the "errors" of measurement $(x_i - a)$, the best value to assume as the result of the experiment is the arithmetical mean.

What right do we have to fall back on the Gaussian distribution function, which is indeed called the "normal error" distribution? Writers on probability have shown that if the errors $(x_i - a)$ are treated as accidental —a negative $(x_i - a)$ being just as likely as a positive one of the same magnitude—so that

$$\sum_{i=1}^{n} (x_i - a) \, P \, (x_i - a) = 0 , \qquad \text{(Equation 14)}$$

then the law of distribution of the errors must be the Gaussian law.[20] This is to be sure an arbitrary definition of accidental errors, but it is plausible.

Other types of distribution functions, corresponding to different hypothetical kinds of errors, are of course possible. The subject of precision of measurement is obviously a large and complicated one. But enough has been said here to emphasize both the important role of probability in measurement and the essentially arbitrary character of the assumptions used in specific cases. Here again the physicist takes refuge in conventionalism in order to function effectively at all. Such validation of his procedures as he can hope to find comes in the consistency between the results of measurements of what is assumed to be the same quantity by different methods in different laboratories at different times carried out by investigators of comparable skill.

The Idealistic Epistemology of Eddington

Insofar as physics is treated as a science whose task is to describe, create, and strive to understand portions of human experience, the physicist is neutral with respect to conflicting philosophical points of view as to the nature of that experience. Whether the experience he works with is produced by a real universe existing independently of man and containing real objects or whether it is all a figment of the human imagination should not influence the attitude of the physicist toward his task. In either case he deals only with the experience itself. In this sense he is immune to the usual philosophical controversies over realism versus idealism in their various interpretations.

But a neutral position in these controversies is perhaps not so obviously dictated. Certainly the objective character of physics as a science strongly suggests—particularly to experimental physicists—the existence of a real world independent of the observer. The fact that careful observers can ultimately agree on the results of experiments would appear to strengthen the case for believing that there really is "something out there" for the physicist to discover. When we find lines in the spectra of the sun and stars corresponding to those observed in terrestrial spectra, the simplest interpretation is that the same elements are present in those far-distant objects as on the earth, and this persuades many that we are indeed observing a single, real universe. The success in classical experimental physics of the object-instrument idea in connection with physical measurements has also tended to reinforce the conception of a real universe of physical objects. Small wonder that experimentalists tend to be realists in the philosophical sense, even though they may not actually have the slightest ostensible interest in the concerns of professional philosophers.

The theoretical physicist may take a different attitude because he approaches the description and understanding of experience from a subjective point of view. The more often his dream of things-as-they-might-be succeeds in predicting new experience, the more he tends to ascribe power over experience to the human mind; he tends to believe that any experience may be but an image created by the mind, something due ultimately to our innate habits of thought. It is not surprising that theoretical physicists are often attracted to the idealistic viewpoint in philosophy, as was the late Sir Arthur Eddington.

This distinguished astrophysicist and authority on relativity and quantum physics reached the conclusion that the universe of experience, in so far as we can comprehend it, is the creation of man's mind. The logic of Eddington's idealistic epistemology proceeds somewhat as follows: What we call our experience is a direct function of our modes of thought, the workings of our minds. We interpret things in terms of space and time, for example, because that is the way our minds function. We even sense things the way we do because of the peculiar sense organs we have (they might conceivably be different), and these sense organs are connected with the nervous system, where we believe thought takes place. It therefore follows that we ought to be able to construct out of our heads the world of experience, even if we never experience it directly. Eddington puts it this way:

Unless the structure of the nucleus has a surprise in store for us, the conclusion seems plain—there is nothing in the whole system of laws of physics that cannot be deduced unambiguously from epistemological considerations. An intelligence unacquainted with our universe, but acquainted with the system of thought by which the human mind interprets to itself the content of its sensory experience, should be able to attain all the knowledge of physics that we have attained by experiment. He would not deduce the particular events and objects of our experience, but he would deduce the generalizations we have based upon them. For example, he would infer the existence and properties of radium, but not the dimensions of the earth.

The mind which tried simultaneously to apprehend the complexity of the universe would be overwhelmed. Experience must be dealt with in bits; then a system must be devised for reconnecting the bits, and so on—. In the end what we comprehend about the universe is precisely that which we put into the universe to make it comprehensible.[21]

To a strict empiricist, whether he believes in philosophical realism or not, this probably sounds like nonsense. However, Eddington took it seriously enough to spend many years in devising a theory to support it. His aim was to calculate the two important non-dimensional constants of nature conceived according to the quantum and atomic theories: (1) the ratio μ of the mass of the proton to the mass of the electron (of the order of 1836), and (2) the fine structure constant $1/a = hc/2\pi e^2$, where h is Planck's constant, c the velocity of light in free space, and e the charge on the electron. This universal constant $1/a$, which has the approximate value of 137, occurs in the theory of the spectrum of hydrogen. From these universal constants and the observed values of the velocity of light in free space, the Rydberg constant of spectroscopy, and the Faraday, it is possible to calculate all the well-known atomic constants, like the mass and charge of the electron. Clearly the success of a program like Eddington's for the theoretical determination of μ and $1/a$ would augur well for the value of his point of view. In his last and posthumous book (1946), *Fundamental Theory*, Eddington actually carried out the proposed calculations and obtained results in rather good agreement with the experimental values accepted at the time. It must be confessed that his methods have not been clear to most physicists, and his analysis must still be subjected to further scrutiny, assuming his idea continues to provoke interest.

Eddington's view is persuasive but is open to the criticism that he has developed a most general idealistic epistemology by utilizing the special results of relativity and quantum mechanics. A radical change in these theories—which future investigation might well bring about—could substantially alter his program. It is difficult to see how a theory general enough to substantiate Eddington's epistemological viewpoint could yield the specific results needed to give some assurance of his viewpoint's practical value without utilizing the results of a theory special enough to be replaced in the near future.

The very multiplicity of theories with which the history of physics presents us is a serious objection to Eddington's view. How shall we ever know which is going to turn out to be the right theory? Moreover, what assurance do we have that the "system of thought by which the human mind interprets to itself the content of its sensory experience" has reached its final stage of development? If it has not, what good does it do to examine its present state in order to infer human knowledge of physics? To this the supporter of Eddington might reply that in each stage in the development of the human mind there will be a corresponding ability to infer logically the kind of experience that people at that stage of development will have. One cannot quarrel with the logic of this, but pragmatically it suggests that in trying to carry out the details at any given stage, one can do much hard work to no particular ultimate purpose. No matter what values Eddington obtained theoretically for the fundamental constants of nature, it is practically certain that more refined experiments will ultimately yield results that differ from his by more than the commonly allowed experimental error. In other words, no matter how attractive or persuasive his idea may be as a general program, its value for all practical purposes reduces to that of any other physical theory.

Any thoroughgoing idealistic epistemology like that of Eddington, if taken too seriously, may serve to introduce undue dogmatism into science. It may tend to impose a priori synthetic judgments on the method of scientific theorizing. The view of Kant and his contemporaries that the human mind is forced to look upon space through the screen of Euclidean geometry probably did no useful service to the progress of science in his day. Even if we are persuaded that all human knowledge about experience is created by the mind, it seems more helpful ultimately to adopt the attitude that we are not yet sure precisely how the mind performs this creation and what influence the cumulative experience of the race has on its activity. It seems reasonable to suppose that as the creation of experience multiplies, the methods the mind uses to cope with it will change. It cannot be considered otherwise than dogmatic to insist that the physics of a thousand years from now will be developed along the lines of present-day physics or to make assumptions as to what part of our current method of theorizing will survive.

The foregoing may be a rather extreme way of attacking a viewpoint like Eddington's. After all, it can be maintained that all physical theories are susceptible to dogmatism unless the purely hypothetical character of the principles involved is continually kept in mind and the possibility of alternative hypotheses freely admitted. A dogmatic attitude may sometimes be salutary by encouraging the adherents of a theory to take it seriously; and science cannot be expected to progress unless theories are taken seriously. Certain fundamental tendencies that have appeared to crop up in the development of all physical theories to date make the whole problem one of the greatest complexity. Why, for example, do we persist in employing the concept of number in attempting to describe experience, and why, in doing so, do we insist on applying measuring devices such as sticks with scales? Does this not imply some intrinsic a priori concept imposed on the mind? And does not the use of instruments based on these choices dictate our whole method of handling experience? While the force of these questions may be conceded, it may still be contended that the way in which the mind approaches experience is the result of a long developmental process that is still going on and therefore can lead in finite time to decided changes in our methods of experimenting and theorizing. In our mental processes we may be subject to certain fundamental limitations or boundary conditions of which we are not completely aware; but the very flexibility of thought that has gone into the creation of the multitude of theories that ornament the forward progress of physics suggests that man is still comparatively free in the way he looks at experience. This very freedom, indeed, often enables him to force his own ideas on experience, to comprehend it in the light of his own conventions; and in this sense Eddington's views are not too different from the views of Poincaré, the out-and-out conventionalist in matters of physical theorizing. However, Eddington believed that the most important part of experience could be completely described and explained by a single theory created by the mind of man;

Eddington was convinced that he had found the theory in the framework of relativity and quantum physics. Poincaré, on the other hand, believed that there are an infinite number of theories that can adequately describe and explain every branch of experience. The choice of theory then becomes arbitrary, though simplicity (or parsimony) and the relative ease of making identification between the concepts of the theory and operations in the laboratory will usually dictate the choice.

Poincaré has been criticized for his conventionalism because it seems to reduce physics to a mere matter of convenience. According to Poincaré, we invent and use those ideas that seem to work, fully realizing that there are always other points of view that will also work if we try hard enough to fit them to experience. This pragmatic approach displeases those who wish to describe an objective world of experience in terms of physical theory. From the discussion of this in Chapter Two it should be clear that the point of view of the author of this book comes closer to Poincaré's than to that of either the confirmed realist or the idealistically minded Eddington. The choice of theories adopted to describe and explain a given domain of physical experience is not, however, a matter of indifference. When a theory is able to subsume a large domain of experimental results and to predict much experience hitherto unknown, no one but a thoroughgoing sceptic can fail to take it seriously, even if he realizes that sooner or later it will show flaws in the face of the relentless growth of the very knowledge it has served to stimulate. The fact that all theories, like all things invented by man—like man himself, indeed—go through a process of growth and decay should not prevent us from paying tribute to the enthusiasm of the inventor of a theory that somehow seems to fit particularly well and to attract eager disciples who are willing to devote effort to its further exploitation.

An interesting suggestion of a possible line of physical research has recently been presented to put in a somewhat new light the problem of the working physicist's neutrality in the conflict between the philosophical theories of a real and an ideal universe. The suggestion is that one of the tasks of the relatively new radio astronomy shall be to search systematically for signals from outer space that indicate the existence of intelligent life there. The question of the possible existence of life outside the earth, in other planets of the solar system or in similar bodies connected with other stars in our own or far-distant galaxies, has long been a fascinating one. It now seems certain that electromagnetic radiation in the microwave range, which radio astronomy has made it possible to detect and analyze, has its origin in distant sources. This has suggested strongly to some scientists the possibility that some of this radiation may well be signals that sentient beings are transmitting in our direction and possibly other directions in an attempt to communicate with us and other living things capable of attaching meaning to such signals.

This imaginative approach has attracted considerable attention. It appears to have a connection with the problem of the physicist's alleged fundamental indifference to philosophical realism or idealism, for any one who seriously believes that intelligent living beings out in space are trying to communicate with us by means of radio signals must also believe that these people have discovered electromagnetic radiation and are using it in a sophisticated fashion. In other words, they must have had essentially our experience. But to assume the existence of such beings in outer space with experience like ours is to assume that there is one real objective world that produces these people's experience and ours alike. Otherwise—if our universe were simply created by us out of our experience in order to provide meaning for this experience—what right would we have to assume that those hypothetical beings out in space, whose existence is not part of our experience save by the very communication hypothesis in question, actually have our experience and invent from it the same kind of world that we invent? Of course it may be argued that a scientist has a right to assume anything he likes as long as it does not involve a logical contradiction. But this kind of assumption seems farfetched and scientifically implausible. At the very least it seems far more plausible to assume that there is a real world and that no matter where intelligent beings exist, they will recognize this real world as we do—that is, if we are prepared to take seriously the possibility of communicating with intelligent extraterrestrial beings. It seems that here we have a much more definite and decisive criterion for holding a particular philosophical belief in the face of the normal neutralism of science than any hitherto presented.

As for the probable value of the search for meaning—our meaning—in the extragalactic signals, no matter what these signals are, it is fairly certain that some clever investigator will ultimately attach some significance to them. Clever scientists are always finding meaning in the experience of our race that we may legitimately call nature's communication to us. For example, ultimately someone is sure to find, by grouping extragalactic signals in proper fashion, that counting is going on in outer space. Some sceptics may question the value of investing great amounts of money in such an investigation, but the wise procedure in such matters is to maintain an open mind and be as generous as possible.

The Meaning of Simplicity

The scholastic principle of parsimony (Occam's razor), treated briefly in Chapter Two, continued to develop throughout the history of physics into the general idea of simplicity as it applies to physical theory. The notion of simplicity is appealing yet elusive. Everyone thinks he knows what simplicity means but is unable to define it precisely even in operational terms. To Newton nature seemed simple, but the nuclear physicist who reads his *Principia Mathematica* will wonder just what he meant. Yet philosophers pay a great deal of attention to simplicity, believing, perhaps, that something more is needed than the agreement of a physical theory with experience—or even a theory's prediction of hitherto unknown experience—to warrant its being taken seriously as a model of the world. So many different theories may satisfy the same set of empirical facts that other criteria are desirable and necessary for the acceptance of a given theory. Simplicity has been one such criterion, and it has long been an attractive one.

The first impression of any natural phenomenon is almost always one of disorder and confusion. Even with our modern sophisticated point of view, much new experience in physics looks as woefully chaotic as early primitive experience. But eventually some clever investigator comes along and by what looks like a magical idea transforms chaos into order. The ingenious theorist manages to seize upon certain aspects of physical experience that appear to be particularly relevant to the physical problem that interests him. To cite a very old example, the Greeks must for long ages have contemplated the behavior of solid bodies immersed in, or floating on, fluids before Archimedes finally hit on the really pertinent facts and as a result established and enunciated the laws of hydrostatics. He finally asked the "right" questions and hence was able to come up with a meaningful handling of the observations. He did this not merely for the satisfaction of knowing something decisive and important about the buoyant properties of fluids but also for pragmatic reasons—so that the results could be applied with confidence to practical problems. In particular, for example, he solved the problem of the determination of the positions of equilibrium for any floating solid having the shape of the right segment of a paraboloid of revolution and having any arbitrary density. Moreover, he was able to predict the stability of the equilibrium—the ability of the object to right itself when disturbed from its position of equilibrium.

Archimedes introduced simplicity through a single principle that replaced his predecessors' rule-of-thumb procedures, which were hopelessly complicated and unreliable and often resulted in disastrously unstable ships. Such developments in physical theory make it difficult to resist the feeling that they occur because nature is essentially simple if only experience is penetrated with sufficient depth. The conviction therefore grew that the correct or true laws of physics are simple laws in the sense that they are expressible in terms of a handful of symbolic concepts and principles that can be manipulated by a few elementary mathematical operations. From this point of view the truth of laws like those of Boyle and Hooke, for example, is seen to be guaranteed by their particularly simple functional form. This was a common argument during the eighteenth century and the early part of the nineteenth. But it is less convincing today, for in contemporary experience, when we probe more deeply into any phenomenon than was possible earlier, the "simple" law or the "elementary" formula no longer seems adequate for the description of the new details. Boyle's law has to give way to the Van der Waals equation or to other more elaborate equations of state of a real gas; Hooke's law must be replaced by something more readily represented by a graph than by a manageable algebraic expression.

The same development runs through the whole of physics. As another illustration, take the familiar falling-body formula ($s = \frac{1}{2}\ gt^2$). It appeared at first sight to insure the essential simplicity of motion near the surface of the earth. But the formula failed to describe accurately the finer details that are revealed by more careful observation. For instance, it did not account for the observed small eastward deflection during fall and for other peculiarities of the motion; these demanded more or less elaborate corrections to the formula. Simplicity seemed to have been lost, but this did not diminish scientific enthusiasm for the concept, and ultimately it was observed that the difficulty is only an apparent one. In our description of the fall we use a system of reference that is rotating because it is fixed in the earth. Further examination showed that the simple formula does hold in a nonrotating frame of reference, as unnatural as it seems to employ such a frame of reference on earth. Here we have exchanged the simplicity of a formula for the simplicity of a more fundamental concept—that of the difference between rotating and nonrotating co-ordinate systems. It appears impossible for us to have our cake and eat it too.

The major difficulty in discussing simplicity evidently is the imprecise meaning of the concept. We might, of course, dismiss the problem by agreeing to define simplicity as merely the assumptions that nature is uniform and that it is possible to utilize the method of science in describing physical phenomena. But it ought to be possible to do more than that. In the introduction to his lectures in theoretical physics, Gustav Robert Kirchhoff, in 1874, defined mechanics as the subject whose task it is to describe completely and in the simplest manner the motions occurring in nature. This definition implies that the success, even the "truth," of mechanics resides largely in its simplicity. But how are we to know what is true? The tendency of Kirchhoff—and, indeed, later physicists—to emphasize the close connection between simplicity and truth has been criticized by Hans Reichenbach, who has introduced two separate notions to which the name simplicity may be attached.[22] The first he calls "descriptive simplicity." It really reduces to the notion of convenience and can best be illustrated by the attitude of the physicist toward two physical theories that describe equally well a given domain of experience. For some purposes one theory may prove more convenient than the other—as when the ray theory of light is more useful in most calculations involving mirrors and lenses than the wave theory, though the latter must be used whenever the phenomena of interference and diffraction enter the picture. We say that for practical purposes in certain applications one theory is simpler than the other; we really mean it is more convenient.

Reichenbach calls the second kind of simplicity "inductive." An example of inductive simplicity is the customary practice in physics of drawing a smooth curve through the discrete set of points resulting from a certain set of measurements. This kind of simplicity is treated by Reichenbach as more fundamental than the purely descriptive variety. He seems to associate inductive simplicity with truth value. According to Reichenbach, the continuous curve has greater probability and therefore greater truth value than the broken line formed by joining the separate points by straight lines. This would seem to associate simplicity with greater probability. A closer look at the two definitions of simplicity shows that under the surface descriptive and inductive simplicity are not really different enough to be validly distinguished. Is there after all definite assurance that the continuous curve drawn among the discrete points describes the observed phenomena more truthfully than the points themselves or the broken line joining one point to another? It is granted that the mathematical technique involved in working with the continuous curve is more familiar to most people than that involved in working with the discontinuous function involved in the discrete points. But does this make the curve more probable or more truthful? It hardly seems so, unless we are prepared to base probability and truthfulness on convenience. This, however, would effectively reduce inductive simplicity to the same level as descriptive simplicity.

What do we really mean by saying that one theory or one physical procedure is more convenient than another? Is convenience perhaps an instinctive concept that is not susceptible to analysis? What has seemed convenient to scientists of one age has seemed otherwise to those of a later period. At one time, many scientists thought it more convenient to describe the properties of matter in terms of continuous substance (caloric for heat and magnetic and electric fluids for the phenomena of electricity and magnetism), whereas others found the atomic view more congenial. To a certain extent the notion of convenience may be equated with the notion of common sense, which has changed decidedly from age to age. Convenience may also be related to the notion of "economy of thought" that was stressed so emphatically by Ernst Mach as a pragmatic criterion of science in general.

If science is a method for the description, creation, and understanding of human experience, simplicity in any branch of science must refer to these three categories. What do we mean by a simple description? Certainly this question would be answered differently by a naturalist and a physicist. The naturalist depicts the structure, behavior, and relations of plants and animals by means of a system of a classification into types; his method is taxonomic. His scheme is of such a character that a careful examination of any particular living specimen (not necessarily alive at the time of examination) enables him to fit it into its proper place in the system. The method of description employs a specially invented technical vocabulary. Presumably the naturalist considers his description simple if it represents the complexity of living things that is apparent to the untutored observer by means of a carefully worked-out classification embodying the smallest possible number of essentially different types.

Physics, however, is not a classificatory science, and moreover its normal vocabulary differs from the taxonomist's terminology in making maximum use of the abstract symbolism of mathematics. One method of deciding what simplicity means in physics is to examine each category in the logical schema of physical theorizing from the standpoint of the idea of simplicity. To do so we must give attention again to the fundamental concepts or constructs, the postulates or hypotheses, the mathematical methods used in the deduction of laws from the postulates, and finally the experimental methods for testing the laws as adequate descriptions of experience.

Of all basic physical concepts, which are the simplest? The physicists of the nineteenth century would have betrayed no hesitation in answering that the simplest are those most closely connected with primitive sense perceptions—space, time, and number, with the attendant properties of the three-dimensionality of space and its Euclidean geometry. It is of course an exaggeration to imply that all our predecessors of the last century felt this way, and it is true that it was difficult to fit a thermodynamic concept like temperature into this pattern, though it seemed instructive for other, mainly biological, reasons. Nevertheless, most physicists of that period believed that a physical theory constructed from concepts based on space, time, and number was a simple theory. Of all physical theories classical mechanics most closely fulfills the assumed criterion, since it operates with concepts most immediately connected with ordinary experience. It is plausible to draw the conclusion from this space-time-number criterion that in order for physics to be simple it should rest on a mechanical basis. Hence the high favor bestowed on mechanical theories during the eighteenth century and most of the nineteenth. Lord Kelvin was so impressed by this point of view that he refused to take seriously any physical theory not based on mechanics. He never accepted willingly the electromagnetic theory of light and persisted in trying to follow out the elastic-solid point of view in optics in the vain hope that it could be made to work.

A similar situation is found in modern quantum mechanics, which has had to introduce concepts quite foreign to those of classical mechanics in order to account for the properties of atoms. Many attempts were made in the latter part of the nineteenth century and first decade of the twentieth to account for atomic spectra, for example, on the basis of classical mechanics. They did not succeed very well, though there seems to be little doubt that if sufficient complexity of detail is introduced, a classical-mechanical description can ultimately be provided not only for atomic spectra but for all physical phenomena. However, most physicists have concluded that the cost in terms of elaborateness and cumbersomeness would be prohibitive. Quantum mechanics has the great merit of working, even though its constructs look strange to those brought up in the classical tradition.

There seems little reason for maintaining the view that simplicity resides essentially in classical-mechanical concepts and that therefore those concepts most closely connected with direct experience are necessarily the simplest ones. This is borne out if the meaning of the state of a physical system in classical mechanics is contrasted with its meaning in quantum mechanics. In the case of a system of particles in classical mechanics its state is said to be specified by the positions and momenta of the particles in some reference system. Thus the state, classically conceived, is essentially a set of numbers attached to symbols that directly represent mechanical constructs flowing more or less immediately from primitive space, time, and number sense data. These numbers and the state they specify may, and usually do, alter with the time. Even in cases where the dynamic method is applied to problems in which motion is not directly observed, as in the dynamic theory of electric currents, the state of the system—in this illustration the circuit under consideration—is given by a set of the directly measurable charges and currents. While the classical construct of state is very closely related to direct experience, the meaning of state in quantum mechanics is more abstract, being specified by a mathematical function of certain variables—space and time co-ordinates or transforms of these. The function itself is not a mechanical quantity in the usual sense but is a kind of source function from which mechanical quantities can be calculated by means of the fundamental formulas of quantum mechanics.

Although the two concepts of state are on different levels of abstraction, who is to say which is the simpler? Many classically educated physicists would naturally prefer the classical construct. However, since anyone with a grasp of the mathematics needed for classical mechanics can understand the quantum-mechanical construct well enough to use it with success in a pragmatic sense, a preference for classical mechanics based on the assumption of its a priori simplicity is unfounded. It is unlikely that a valid criterion can be found for simplicity in the fundamental constructs of physical theories. Still, every theory rests on a plurality of constructs. Could we not plausibly say that the simplicity of a theory is measured inversely by the number of independent constructs necessary for its establishment and expression?

A striking illustration of the plausibility of such a judgment is provided by the problem of the nature of light in a comparison of the corpuscular theory of Newton with the wave theory of Huygens and Fresnel. (We have already examined the distinctions in these theories in some detail in Chapter Two in connection with the place and nature of the so-called crucial experiment in physics.) The light corpuscle of Newton was undoubtedly suggested by the concept of the material particle of classical mechanics; it would have been a satisfactory concept if the phenomena of optics could be described merely by endowing the corpuscles with extrapolations of the ordinary properties of such a particle—vanishingly small inertia and size, very high speed, and the power of interaction with material particles. Unfortunately, Newton found it necessary to attribute to the corpuscle additional properties very different from those of observed material particles.

He had to suppose, for example, that the corpuscle, when incident from a vacuum on the surface of a body, is attracted to the body by a force that acts only perpendicular to the surface. Moreover, in order to account for the color phenomena associated with the incidence of white light on thin transparent plates, he had to assume the possibility of the corpuscle entering a kind of transient state that disposed it to be readily transmitted or reflected. Finally, to account for the polarization of light he was forced to postulate that every ray of light has two opposite sides. All these assumptions involve the introduction of what can only be considered independent constructs added to and foreign to the nature of the material particle of classical mechanics.

Greater simplicity appears to characterize the wave theory of light of Huygens and Fresnel, the theory based on the assumption that light is propagated by waves like the elastic waves observed in the ordinary solids of our experience. The wave concept suffices—without the addition of gratuitous postulates—to describe not only reflection and refraction but also the interference, diffraction, and polarization of light. Difficulties arise when the properties of the medium are examined closely: the "elastic solid" constituting the light-bearing medium has to be more elastic than steel and yet so tenuous as to oppose no observable resistance to the motion of the planets. This theory errs as the corpuscular theory does by invoking extraconceptual properties not contemplated in the classical theory of material media. However, this difficulty was obviated by the reminder of Sir George Stokes that even certain ordinary substances like pitch are comparatively resistant to rapid deformation but offer little resistance to slow deformation. The construct of the elastic-solid luminiferous medium would thus appear to be a reasonable extrapolation of the elastic solid of ordinary experience.

A more serious problem is presented by the necessity of accounting for the absence of the longitudinal wave that exists in disturbed elastic solids but is not observed in optical phenomena. This problem was finally solved only at the cost of considerable elaboration, which most contemporary physicists would regard as a sacrifice of simplicity. Ironically enough, by the time the problem had been solved, the elastic-solid theory of light had been shelved effectively and replaced by the electromagnetic theory. This brought to the fore a new construct, that of the electromagnetic field. It proved highly successful, though it also introduced serious problems related to the nature and behavior of the medium through which the light radiation passes. However, like the elastic-solid theory, the theory of the electromagnetic field satisfied the criterion of simplicity more successfully than the corpuscular theory of Newton.

Another illustration of the "minimal-construct" or numerical criterion of physical simplicity is provided by a comparison of the eighteenth-century caloric theory of heat with the mechanical theory of heat. It is clear that the caloric theory falls short of simplicity by being forced to the uneconomical expedient of introducing an unnecessary concept, sharing with ordinary matter its indestructibility, but being unlike it in such other respects as weightlessness. Thus it has to enter as an *ad hoc* independent construct. The mechanical theory of heat, on the other hand, is under no such necessity; here the principle of conservation of energy is sufficient.

In examining these examples of the minimal construct criterion, a fundamental question arises: Just what do we mean by the term independent construct? In the cases of the corpuscular theory of light and the caloric theory of heat the matter seems reasonably definite. But when the term independent construct is applied to the constructs of classical mechanics (material particle, displacement, velocity, acceleration, force, mass, and energy), the question of whether they really are independent must be considered.

In mechanics a material particle has all the properties of a geometrical point in Euclidean three-dimensional space. It also has two other properties of a nongeometrical nature: (1) certain relations to other particles, so that its motion is in general dependent on its position relative to other particles, and (2) inertia, the measure of which is its mass. These concepts become precise only when described in terms of the kinematical concepts of displacement, velocity, and acceleration.[23] But displacement and velocity are mutually related. If time and space are both necessary fundamental elements in which mechanical description is carried out, then all displacement must take place in time, and the concept of velocity is not independent of, or unrelated to, that of displacement.

It is idle to argue that the precise symbolic form in which velocity is expressed (whether distance divided by time or time divided by distance) represents a choice not dependent on the concept of displacement. Once motion is admitted to be something that takes place in both space and time, the concept of displacement is logically unthinkable without a corresponding concept relating displacement with time, even though the precise manner of defining this corresponding concept is a matter of indifference. The same arguments must hold for acceleration and for all higher time derivatives of displacement that might be introduced into mechanics.

The assumption that the material particle has certain relations to other particles is an essentially new property compared with those of position and change of position alone, and it is nongeometrical. Mechanics conceivably might operate with swarms of particles having no relation to each other. (For example, only the number per unit volume might be significant.) Actually, mechanics does not operate this way, and the precise assumptions underlying the definition of mass are also independent of the purely kinematic concepts.[24] On the other hand, on the basis of the foundations of classical mechanics, the construct of force cannot be considered as independent of the constructs of kinematics, the assumed interrelations of particles and mass.

The whole of mechanics is perfectly conceivable without explicit reference to force—as in Karl Friedrich Gauss's principle of least constraint and Heinrich Rudolph Hertz's development of the theory of mechanics. It is also perfectly conceivable without explicit reference to the construct of energy. These concepts arise not so much through new assumptions as through the introduction of certain combinations of symbols that occur frequently and therefore seem to merit special attention and use. This is quite a different thing from independence in the usual sense. Naturally the viewpoint taken here excludes the introduction of force and energy as anthropomorphic constructs wholly outside the purely logical development of mechanics.

Any classical-mechanical theory of a realm of physical phenomena contains a definite number of independent constructs. Any additional assumptions about material particles transcending the fundamental principles of mechanics—such as that particles can be created out of nothing or that they can be destroyed—are foreign, independent, and unrelated concepts. On the other hand, the usual initial boundary conditions of mechanical problems appear merely as specifications suitable for defining the solution of the equations of motion for a particular problem. The boundary conditions cannot be considered as entailing anything independent of the fundamental constructs. The choice of the Hamiltonian function for a dynamic system involves the same kind of specification. The principles of mechanics lay down no rules for the choice of such functions beyond the usual purely analytical requirements, such as those of differentiability. The choice of a particular Hamiltonian can therefore not be considered an unrelated assumption grafted on to the fundamentals.

It is often stated that all physical phenomena can ultimately be reduced to a description in terms of classical mechanics. If this were the case, the number of fundamentally independent, unrelated constructs would be the same for all such descriptions, and on the numerical criterion of simplicity they would all be equally simple. Few physicists would be willing to accept this as a sensible conclusion, however, and therefore what we have called the minimal-construct criterion of simplicity becomes illusory unless we deny that all physical phenomena can be reduced to a description in terms of classical mechanics. If we do deny that it is possible to provide a genuine classical-mechanical description of atomic phenomena, and looking for a completely different construction, find it, let us say, in quantum mechanics, we are begging the question as far as a decision on the numerical criterion of simplicity is concerned. And if we accept the criterion, we are faced with the fact that quantum mechanics operates with the constructs of classical mechanics but adds to them the wholly foreign concept of the ψ function and the reinterpretation of the construct of state. On this basis we must confess that it is necessary to forgo the simplicity of classical mechanics for the sake of describing new and more complicated phenomena. A quantum physicist would probably prefer to conclude that the numerical criterion of simplicity is meaningless and that simplicity cannot be measured but must be felt intuitively—in one's bones, as it were.

The numerical criterion of simplicity encounters still another difficulty: it takes no account of the relatively great difference in nature between dif-

ferent constructs. Two theories may operate with precisely the same number of constructs, yet those of one may be relatively much more abstract
than those of the other. For example, the quantum-mechanical concept of
state is more abstract than the concept in classical mechanics, in the sense
that the former has less relation to actual human experience than the latter. It seems futile to endeavor to compare such radically different ideas
from the standpoint of simplicity.

There is, we must conclude, little or no satisfaction in attempting to
apply the numerical-simplicity criterion in the constructs of physics. The
process of physical reasoning includes not only the constructs but also the
postulates that relate the constructs and the mathematical methods of
manipulating the postulates in the deduction of physical laws. Since the
postulates in a very real sense define the meaning of the constructs, most of
what has already been said about simplicity in connection with constructs
applies equally well to simplicity in connection with postulates. How, for
example, is one to compare decisively the assumptions of the kinetic theory
of gases with the principles of thermodynamics? The former are so specific
and the latter are so general that they appear to be of an entirely different
character; any merely numerical comparison appears to be utterly out of
the question.

The same inference is even more plainly indicated by an examination of
the mathematical analysis used in expressing physical theories. It is perhaps more properly the province of the mathematician to speak on simplicity in mathematics. Nevertheless, it must strike most physicists that in
the progress of mathematics from algebra and geometry to higher analysis,
new constructs have been steadily added. Whatever gain there has been in
simplicity has been in the direction of economy and utility, for the sake of
which the scientist has been willing to pay the price of increasing abstractness.

If the concept of simplicity in physical theory cannot be defined in
strictly numerical terms, can some other basis be found for it? One possibility is the "economy of thought" stressed by Mach as one of the great
contributions of scientific method to human activity. The laws and principles of contemporary physical science make possible the intellectual
mastery of nature to an extent inconceivable to the mind accustomed to
look upon each natural event as unrelated to every other one; it makes it
possible within a year to teach an intelligent, industrious college student
to solve with ease problems that would have given considerable difficulty

to Aristotle and would very likely have been solved incorrectly by him—if he could have solved them at all. The tremendous practical applications of science all attest the marvelous economy of mental effort inherent in the method of physics. It is as if in the process of its development the human mind were always striving to achieve an understanding of natural phenomena in the shortest possible time and with the least mental effort—trying to find a way that does not involve retaining in the memory a large number of isolated facts that might be replaced by a few great generalizations.

Mach's economy of thought provides a pragmatic criterion for simplicity in physics that may not turn out to be illusory: of two theories descriptive of the same range of physical experience, the simpler one is that which demands the shorter time for the normally intelligent person to become sufficiently familiar with it to obtain correct and useful results. No question is here involved of the precise type of constructs employed, their number, the number of fundamental postulates, or the special mathematical schemes employed in making deductions. If, for example, a person familiar with classical mechanics can become equally well acquainted with another physical theory in the same length of time that he took to learn mechanics, he should consider the new theory as simple as mechanics, no matter how complicated it may seem on first examination. Human life is short and time is fleeting. In endeavoring to understand the world around us, we must make the most of the brief span allotted to us. We must therefore build our theories in such a way that their manipulation will lead in a minimum length of time to success in physical prediction.

This disposition of the meaning of simplicity in physics suggests that because of individual differences among scientifically minded people simplicity must ever remain essentially an individual judgment—a theory that is simple to one will not appear so to another. Nevertheless, past experience shows that gradually, as a matter of successive approximations, points of view have arisen that are more readily learned than earlier ones. These then have become part of the more or less stable and at least quasi-permanent body of physical thought. As new theories arise for the description of newly acquired phenomena, they too will take their proper place. And in turn we shall ultimately consider them simple when we have grown sufficiently familiar with them to forget that they ever seemed difficult to understand.

Another aspect of the problem of simplicity in physics is the nature of physical experimentation. We have already seen that the progress of a science like physics is inextricably associated with the creation of new experience; consequently, simplicity would seem to have a bearing on experimentation. If, going back to the previous discussion of simplicity in physical theories, we analyze the various meanings to be attributed to simplicity in experimentation, we find them closely analogous to our assignment of this character to constructs and postulates. Perhaps those experimental designs and equipment are simplest that are most closely connected to the sorts of things we observe in direct experience. The balance being a device of the greatest antiquity, it would be natural to employ it for scientific purposes in comparing masses or weights of different objects and in the measurement of specific gravity. On the other hand, an electron accelerator would be considered a departure from simplicity, for it bears little relation to devices available to previous technology.

As we explore this problem, however, we encounter all the difficulties experienced in the earlier discussion of simplicity in connection with physical theories. We shall probably be led to the same pragmatic conclusion reached there, that simplicity in experimentation is largely a matter of familiarity. The method and apparatus in question must, of course, create experience that lends itself to scientific description and understanding. Once this is admitted the rest is largely a matter of learning how to follow the recipe of the inventor. This does not preclude the search for alternative ways of measuring a given physical quantity—the charge on the electron, for example. Some of these alternatives may involve less expensive equipment, cheaper materials, and less time in fabrication. In a pragmatic sense we may therefore call them simpler, and this may have an effect on the choice of material in teaching physics.

The Relation between Mathematics and Physics

Physicists use mathematics in their attempt to describe and understand experience. In place of the language of ordinary speech they rely heavily on abstract symbols and express relations among them in terms of other symbols that represent operations like addition, subtraction, multiplication, division, differentiation, and integration. As we have stressed in connection with the concept of measurement, physics is quantitative in the sense that it ever seeks to answer the question "how much" rather than merely "how." And since the expression of quantity requires numbers, physicists use mathematics to deal with the numbers and the various operations that may be performed on them.

This is by no means the whole story, however. Mathematics provides a simple means of expressing functional relations between symbols to which operational significance can be attached, and from such relations logical deductions can be drawn by mathematical manipulation. The logical deductions in turn can be identified with the results of laboratory operations or other observations, and a wealth of descriptive material can thus be amassed in a relatively economical and accurate fashion. This aspect of the use of mathematics in physics—the combination of logical deductions by means of mathematical analysis with laboratory experiments—implies the desirability of creating mathematics adequate for the needs of physical description and explanation. It also requires physicists to search for ways to utilize in physics the powerful new mathematical techniques invented by pure mathematicians. The history of physics and mathematics provides many examples of both lines of endeavor.

For a statement on the essential role of mathematics in physics we can hardly do better than quote from the address delivered by James Clerk Maxwell before the British Association for the Advancement of Science at Liverpool in 1870:

There are men who, when any relation or law, however complex, is put before them in a symbolical form, can grasp its full meaning as a relation among abstract quantities. Such men sometimes treat with indifference the further statement that quantities actually exist in nature which fulfill this relation. The mental image of the concrete reality seems rather to disturb than to assist their contemplations.

But the great majority of mankind are utterly unable, without long training, to retain in their minds the unembodied symbols of the pure mathematician, so that, if science is ever to become popular, and yet remain scientific, it must be by a profound study and a copious application of the mathematical classification of quantities which lies at the root of every truly scientific illustration.

There are, as I have said, some minds which can go on contemplating with satisfaction pure quantities presented to the eye by symbols, and to the mind in a form which none but mathematicians can conceive.

There are others who feel more enjoyment in following geometrical forms, which they draw on paper, or build in the empty space before them.

Others again, are not content unless they can project their whole physical energies into the scene which they conjure up. They learn at what rate the planets rush through space, and they experience a delightful feeling of exhilaration. They calculate the forces with which the heavenly bodies pull at one another, and they feel their own muscles straining with the effort.

To such men momentum, energy, mass are not mere abstract expressions of the results of scientific inquiry. They are words of power, which stir their souls like the memories of childhood.

For the sake of persons of different types, scientific truth should be presented in different forms, and should be regarded as equally scientific, whether it appears in the robust form and the vivid colouring of a physical illustration, or in the tenuity and paleness of a symbolical expression.[25]

Maxwell felt strongly the need for mathematics in the development of physics, though he recognized that in the presentation of physical ideas to the inquiring nonphysicist it might be necessary to translate the content of the mathematics into ordinary language. In considering this problem, it is necessary to keep in mind that nowadays there are two kinds of mathematics and two kinds of mathematicians—the pure and the applied. Bertrand Russell defined pure mathematics as the class of all propositions p implies q, a definition that effectively reduces mathematics to deductive logic. Russell's definition commands great respect by emphasizing a cardinal virtue of mathematics—indeed, the one that impresses most people, even those who feel a repugnance toward the use of it—that in essence mathematics is really nothing but an application of logical thinking.[26] Although certain logical paradoxes somewhat dim admiration for Russell's definition, the physicist is not apt to be too seriously worried about them. To the physicist mathematics is, as J. Willard Gibbs is supposed to have called it, a language. It is probably the most remarkable and powerful language ever developed by the mind of man. As a language, it must be painstakingly learned, like any other; the reason many people claim to be unable to understand mathematics is not that they are incapable of thinking logically but simply that they have not troubled to learn the language.

The pure mathematician is not content to limit his activities to those of a linguist, although he undoubtedly feels that there is more clarity in mathematics than in verbal language. The pure mathematician would like to view mathematics itself as a science like physics or chemistry, and it is not too difficult to accept this assessment. Except for the category of experimental verification, pure mathematics can be analyzed logically in precisely the same way that a physical theory was analyzed in Chapter Two. Since the counters or entities that form the working elements of a mathematical theory have no identification with anything in the world of our ordinary experience, the only test of the value of a pure-mathematical theory would appear to be in the establishment of its internal consistency. It may be objected that in a branch of mathematics like geometry an experimental test of the theorems (the deduced laws of the geometrical theory) can be conducted by drawing appropriate diagrams and measuring them with ruler and protractor; but the counterobjection is that when so conceived, geometry ceases to be pure mathematics and becomes a branch of physics. If, on the other hand, we are dealing with a branch of pure mathematics in which the deduced results are testable by counting—a large segment of mathematics—we are obviously permitted to look upon mathematics as a science subject to experimental verification, for counting is an experimental technique whether done with the fingers or by the use of a sophisticated computing machine.

If mathematics is essentially the expression of logical relations through symbols, its value for physics in terms of economy and power is almost self-evident, since in physics we clearly use deductive logic in all inferences. Those who are alarmed by what they believe is the unnecessary abstractness of mathematical symbolism may find some comfort in what Maxwell had to say about Michael Faraday in the Preface to the first (1873) edition of Maxwell's famous *Treatise on Electricity and Magnetism*. Maxwell tells us that he began the composition of this work with the supposition, common at that time, that there was a decided difference between Faraday's way of looking at electrical and magnetic phenomena and that of the continental European school of mathematicians. However, as he proceeded with his study of Faraday's *Experimental Researches in Electricity* (1882), Maxwell became convinced that Faraday's method of description was also a mathematical one, even if it was not expressed in the conventional mathematical notation of the time. We now credit Faraday with originating the field concept in electrical science; Maxwell's great contribution was to translate Faraday's theory into the accepted mathematical terminology of the nineteenth century. In the use of his geometry of lines of force Faraday reasoned mathematically in a very decisive fashion. It is not at all unlikely that the more powerful modes of nonmetrical mathematical thinking typified by modern topology may ultimately find as useful an application to modern physics as Faraday's geometrical notions.

Maxwell's mathematical translation of Faraday's theory illustrates an important aspect of the use of mathematics in physics, the possibility of employing quite different types of mathematical treatment for the same physical problem and the converse possibility of using the same mathematical technique in the description of a wide variety of different physical situations. The latter is, of course, the basis for the successful introduction of analogies in physics, which has been discussed in Chapter Two. But the former deserves some serious attention, for even in the study of pure mathematics it is much more illuminating to solve a given problem by two or more different methods than to solve any number of problems by the same method. As a pedagogical procedure this is not always accepted with enthusiasm by students of the subject, but it is an idea that has led to tremendous strides in physics as well as in mathematics. For example, it is perfectly possible to develop the theory of the behavior of elastic media in terms of sets of simultaneous equations connecting stress components with strain components; but much more insight into, and manipulative power over, these problems is gained if we translate the results at once into the notation of tensor calculus. Maxwell stressed this point of view when he said that mathematics is a method for saying the same thing in many different ways. Other illustrations come to mind, among them that of quantum mechanics, which can be developed by the use of the matrix analysis preferred by Heisenberg or by the operator method used by Schrödinger. Each development has its peculiar advantages; the end results are identical. A further illustration is the variety of mathematical techniques that have been devised for building the theory of mechanics—Newton's laws versus the method of Sir William Rowan Hamilton and Joseph Louis Lagrange, which is based on making a certain integral take on a stationary value.

The search for new techniques constitutes a large part of the subject matter of applied mathematics. Often these schemes emerge directly from branches of pure mathematics originally cultivated from curiosity and the joy of invention with no thought of practical application. Imaginary and complex numbers, for example—originally mathematical curiosities—have proved to be immensely valuable in hydrodynamics, electric circuit theory, and wave propagation. In the mid-nineteenth century when Arthur Cayley first concerned himself with matrices, they were considered a purely academic branch of higher algebra. Now there is scarcely a field of physics in which matrices do not find a use. Similarly, integral equations seem precisely made to handle problems in hereditary mechanics like those encountered in the plastic deformation of solids; and without Fourier integrals and transforms the analysis of complicated wave forms would be impractical. Tensors, when first discussed by mathematicians, seemed to be only a kind of generalized vector that was wholly unnecessary in physics, for which the simple three-dimensional variety seemed adequate; nevertheless, tensors turned out to be precisely what Einstein needed to apply his general theory of relativity to the problem of gravitation.

The history of mathematics and physics shows conclusively that the pure mathematics of one period will be applied to physical problems in the next. There is no reason to doubt that this sequence will continue to prevail in the future and that the pure mathematics of today contains within it the language of tomorrow's physics. The pure mathematician, like G. H. Hardy, may not care to contemplate this. What Hardy calls "real" mathematics must be justified as art, if it can be justified at all; it has no other defense. Moreover, he would not have it otherwise "useful" to anyone. The thought of such possible utility would destroy his joy in the subject.[27] This is an extreme view, probably not shared by many other pure mathematicians. Even if it were, it would make no difference. Physicists will continue to cast about for the abstract symbolism they need wherever they can find it.

It is, of course, commonly known that much of the mathematics used in classical physics and even some of the mathematics that forms the scaffolding of modern physics was invented by physicists themselves. Newton's invention of the calculus is a commonly cited example. In more recent times J. Willard Gibbs and his development of vector analysis come to mind. There may be some argument over whether Gibbs's was an independent invention of a new branch of mathematics or the introduction of an extraordinarily useful notation. No one questions, however, that Oliver Heaviside's operational analysis for the solution of electric circuit and wave propagation problems was a new invention in mathematics. In fact it was so new that mathematicians objected seriously to it because of its alleged lack of rigor. This did not seem to bother Heaviside. He knew his work was pertinent and that the mathematical rigor would ultimately be supplied, as indeed it was by the theory of the Laplace transform. Although it is uncertain whether Fourier, D'Alembert, the Bernoullis, Euler, Laplace, and Lagrange—the great mechanists of the eighteenth century—are to be categorized as physicists or mathematicians, there is no question that among their other activities they constructed mathematical techniques for the solution of problems in physics. And this process goes on today, as in Dirac's formulation of the theory of quantum mechanics; his delta function is so discontinuous that it at first shocked the pure mathematicians, who nevertheless found it worth while later to transform it into rigorous mathematics.

The illustrations from the work of Heaviside and Dirac raise the question of whether physicists have any excuse for using "poor" mathematics.

In general, the pure mathematician will consider bad mathematics those manipulative devices that are merely plausible but have not been subjected to rigorous proof. Yet the physicist continually uses such merely plausible devices. In using infinite series he does not always trouble himself over the problem of their convergence, and he often casually treats discontinuous functions (through differentiation) as if they were continuous. In the early development of quantum mechanics it was assumed that the same theorems that hold for finite matrices will also hold for infinite matrices (those in which the number of elements is unlimited). These mathematical flaws were common among the mathematical physicists of the eighteenth century and even persisted into the twentieth. We have already noted Heaviside's introduction of operator calculus based only on plausibility considerations. To the strictures of the pure mathematician on such practices, the physicist is apt to reply that if the manipulative scheme in question works, it is foolish to fuss over mathematical rigor. The physicist decides whether or not a scheme works by noting the agreement of the results of his analysis with experience in the laboratory. Problems arise, however, when the results of a given theory fail to agree with experience. Are the fundamental physical postulates then at fault, or does the fault reside in questionable mathematical methods? There is no way to settle this matter, as long as doubt remains about the mathematics. From this point of view the logical character of the mathematics used can assume considerable importance.

Pure mathematicians like to poke fun at the attitude of the physicist that in the last analysis all conclusions are tested and either verified or refuted by experiment. They tell an anecdote about a physicist who proposed to prove by straightforward inspection that all odd numbers are prime. The physicist was able to proceed through 1, 3, 5, and 7, but had trouble with 9. "This," he said, "must be an experimental error, since 11 and 13 satisfy the theorem." This anecdote suggests that it is wise in physics to use the best mathematics available, which is to say that we ought to reason as logically as we can and avoid fallacies by scrutinizing our reasoning as searchingly as possible. Most physicists will probably agree with this, though even on this point some cannot refrain from pointing out that Aristotelian logic—the logic used in the everyday affairs of life and adopted by Euclid and most mathematicians—is not the only possible kind of logic. By its dependence on the excluded middle class (the principle that any entity whatever is either A or not A), it is a two-valued logic. There are indeed multiple-valued logics in which mathematical proofs by the *reductio ad absurdum* method, so favored by Euclid and even more recent mathematicians, become impossible. Some physicists have thought that a three-valued logic would assist in the development of quantum mechanics.

The possibly insoluble problem under consideration here essentially involves what is now technically referred to as information theory, for in the last analysis the question concerns the adequacy of the communication of the content of any physical theory. Communication is a function of the language used and will change from age to age as language is modified in the direction of greater articulatory power. As long as man retains his curiosity about his experience and endeavors to express his interpretation of it in terms of relations among apparently diverse phenomena, he will continue to use mathematical reasoning. This will be true whether he is concerned with metrical or nonmetrical aspects of experience. Physics without mathematics would degenerate into an ineffective pastime.

Chapter Four

EXCURSIONS INTO THE HISTORY OF PHYSICS

A Scientist's View of the History of Physics and Its Problems

A scientist's attitude toward the history of physics is obviously based on his conception of history in general. Various epitomizing definitions of this branch of knowledge have been offered.[1] A satisfactory one is that of William Henry Walsh, according to whom history is "an intelligent reconstruction of the past."[2] His definition emphasizes the two important characteristics of history: the accurate presentation of the record of the past wherever found and the rational interpretation of what took place. While Walsh's definition does not specify the precise meaning of "the past," most professional historians prefer to construe it as meaning the past of the human race. To the scientist, however, it means the past of nature and our universe as a whole.

A difficulty arises from Walsh's point of view at the very outset. Physics as we know it today is a well-established discipline, involving specific types of mental attitudes toward experience and its creation. What elements in the past, especially the remote past, should be recognized as corresponding to the physics of our time? This must be an arbitrary decision. Where the Greeks make statements about the motion of bodies the difficulty is not apparent, and we accept their statements as descriptions of a branch of ancient physics. But when the discourse is about dreams and demons, or about temperament as related to the "heat" of the human body, the relationship to modern physical concepts is not so clear. Perhaps the ancient thinkers used such concepts to help them understand that part of experience now called physical, but it is impossible to be certain in all cases.

This leads to another difficulty. Granted a written record that says something about physical experience, exactly what do the words in the record mean? This is not just a problem of adequate translation from an ancient language to our own. We shall assume that a reliable linguist has done this job satisfactorily. Almost inevitably, however, he will not be a scientist and will be forced to give the words their common meaning. But even in modern physics common terms are given meanings entirely different from those they have in ordinary speech, and this was certainly also the case in antiquity.

The scholar-scientist can of course examine the material carefully for internal clues about the meaning of words and other expressions. In the final analysis, however, he must make some further arbitrary judgments. In essence what he does is to frame a theory about the nature of the historical material he is trying to understand. Theory in history is similar to theory in physics. The fundamental constructs of physics correspond to the terminology of the historical record, and the postulates to hypotheses as to the meaning assigned to historical terminology. The conclusions form an estimate of the success of earlier physicists' methods of coping with experience. The historian of physics endeavors to reconcile the thoughts and methods of the past with the modern way of understanding the experience in question; otherwise the whole study would be meaningless. This is perhaps the principal assumption of all history—that it is possible to reconstruct the past in terms that make it intelligible to the present. It is scarcely sufficient, for example, to find and reproduce Babylonian astronomical tables, so that a modern reader can see merely what they look like. The task of the historian of astronomy is not achieved until he has explained just how these tables were used and the theory by which they were constructed.[3]

The task of the historian of physics is clearly a very difficult one. In addition to being a good physicist, he must also be enough of a historian to be able to deal successfully with the problems of reconstructing the past. Such a combination of talents is rare, and the preparation of scholars specifically in this field is a recent development. In spite of the existence of the classical work of Pierre Duhem on the history of mechanics and the more recent work of Alexandre Koyré on Galileo, it cannot be said that the history of physics has made great advances. The number of professional historians of science has grown very slowly, and the fields covered are limited. Even in the development of physics in the seventeenth and eighteenth centuries, for example, there is much that is obscure, though the publication of the collected writings of the distinguished physicists of that time has been actively fostered. The recent introductions by Clifford Truesdell to the newly published editions of the writings of Leonhard Euler show how little understood are the relations among the great mechanical geniuses of the eighteenth century—the Bernoullis, D'Alembert, Euler, Lagrange, and Laplace. It is hardly to the credit of the English-speaking peoples that a complete edition of the writings of Isaac Newton is only now under way.[4]

There are indications of an increasing contemporary interest in the history of physics. One sign is a growing concern to make the record complete and accessible. Not too much can be done about this for earlier periods, but something can be accomplished for contemporary physics or at any rate the physics of the past seventy-five to one hundred years. It is obvious that if there is no record there can be no history. Not all physicists are historically minded and so do not always appreciate the importance of retaining notes, records of experimental measurements, precise descriptions of equipment, and experimental equipment itself. The successors and heirs of physicists are apt to be even more careless and casual about such matters, so that much material of the greatest value to the historian of physics is destroyed or scattered and effectively lost. To prevent this in the case of physics in the United States, the American Institute of Physics, assisted by a grant from the National Science Foundation, has for the past few years been carrying on a project to gather material useful for the history of American physics in the twentieth century. In addition to locating and encouraging the preservation of manuscript material left by distinguished American physicists now deceased, the project endeavors to build up a body of source material contributed by living physicists and to gather information about the location of the equipment they used in their experiments. The records of research published in the professional physical journals do not tell the full story of the mental processes followed by the investigator; for this his notes are often more useful. Attempts are also being made to interview older physicists and to make tape recordings of their views as to the significance of their work. When used with due caution, this kind of record can be of great historical value in answering questions of priority of discovery and other important questions. At any rate the project should provide a more detailed and accurate record of twentieth-century American physics than can be expected to result from the usual chance accumulation.

It must be admitted that there are competent physicists who gravely doubt the value of such concern with the history of their discipline. Some frankly take the stand that no matter how hard the historian of physics tries to unearth the truth about a particular discovery in physics, it will elude him. These physicists therefore conclude that the whole activity leads only to frustration and to misconceptions of the development of the science. One modern proponent of this view is Samuel Goudsmit of the Brookhaven National Laboratory,[5] and on more than one occasion the late P. W. Bridgman also expressed his skepticism of the value of the history of physics. Of course precisely the same sort of attack can be made on history as a whole.

Even when the practicing physicist does not quarrel with the cultural value of the history of physics for the general public, he is often convinced that it has no practical value for the physicist in his research. This view has been expressed with some force by James Bryant Conant, who is willing to admit that an acquaintance with the science of the past fifty years may be useful, but cannot see any use in going back any further.[6] However, a reasonable interpretation of many lines of development in the history of physics effectively refutes this view.[7] Regardless of the difficulties associated with the history of physics, it is clear that they will be overcome and that the subject will ultimately assume its proper place in the intelligent reconstruction of the past that is the task of history in general.

In the light of the discussion of the relation between physics and philosophy in Chapter Three, what, if any, is the connection between the history and philosophy of physics? It is plausible to hold the view that these two disciplines have no essential connection with each other. One might hold that the history of physics is merely the record of what took place and of who did what and when. This would reduce history essentially to an anecdotal status, and its interest to physicists or other scientists would be minimal. But the record in the case of the history of science is supremely important, just as it is for any branch of history. The record alone, however, is not history: it must be interpreted and related to our present knowledge. Since interpretation inevitably involves philosophical matters of the kind already discussed in Chapter Three, the history of physics, to be of any value, must be inextricably related to its philosophy.

The evolution of the concept of energy—one of the great developments in the history of physics—is an example of this inextricable relationship. The germ of the concept may well have arisen in the idea of invariance, the notion that something stays constant in the midst of change. This brings to mind the definition of the total mechanical energy of a dynamical system of interacting particles as the sum of the total kinetic energy of the system and the potential energy (a function of the mutual positions of the particles); this sum stays constant in time, even though the two kinds of energy may vary by themselves, as the particles speed up or slow down and change their mutual distances. On the other hand, it is possible, and has seemed more plausible to some authorities, to seek the germ of the idea of energy in the presumptive impossibility of perpetual motion. This view was favored by Ernst Mach. It is for the philosophy of physics to decide to what extent these two views can be reconciled, or if they cannot be reconciled, which of the two gives the more basic view of the modern concept of energy. Examples of this kind of philosophical problem occur in all aspects of the history of physics.

A related question is whether, if we cannot effectively study the history of physics without reckoning with philosophical problems, we can hope to study philosophical questions relating to present-day physics without some acquaintance with the historical evolution of the ideas in question. This is a controversial matter. It can be argued that philosophical problems, such as those connected with modern quantum mechanics, can be profitably discussed without reference to the history of the subject. For example, the meaning of measurement in quantum physics is a part of the philosophy of physics, but it can be handled merely as an exercise in logical analysis, depending on the various forms of the theory. It seems, however, that a more meaningful procedure would be to examine the meaning of measurement as the idea has developed historically. From this standpoint a philosophical problem in modern physics profits from a knowledge of history. For all practical purposes the philosophy and history of physics are really inextricably related.

The following "excursions" into the history of physics (in the realms of magnetism, atomic theory, the motion of bodies, and acoustics) are illustrations of the types of problems in the general field of the history and philosophy of physics that have appealed to one scientist. These are not systematic and highly documented historical studies. Rather they reflect the interest of one practicing physicist in certain specific historical problems. These problems puzzled him and led him to examine the available literature and give himself the fun of exercising his own imagination in constructing plausible interpretations. Each of the four excursions that follows is an attempt to develop the historical background of a problem, the biographical details about the physicist concerned, and certain related philosophical problems. The section on acoustics strikes a somewhat different note from the other excursions by endeavoring to give a brief sketch of the salient features of the whole historical development of that branch of physics. The bibliographical references in this section are mainly to original sources, and this provides a unique opportunity for placing the evolution of the science of acoustics clearly in perspective and for stressing the relations between this discipline and other branches of science. To a certain extent the section on acoustics epitomizes the general considerations in the earlier parts of the book.

William Gilbert and Magnetism

This first excursion into the history of physics will summarize some points in the history of magnetism and discuss in more detail the work of the English physician and natural philosopher William Gilbert. Examination of his theory of the relation between magnetic dip and terrestrial latitude will illustrate some of the difficulties encountered by the historian of science in interpreting the ideas of earlier scientists. At the same time, it will suggest what a fascinating subject the history of physics can be.

In order to understand a scientific investigator's contribution, it is necessary first to know something of the previous history of his science as well as the state of contemporary knowledge of his specific subject and the contemporary attitude toward scientific research in general. As with many other natural phenomena, no one knows who first observed magnetism; its discovery is shrouded in the mists of antiquity. But certainly the two fundamental properties of the loadstone, or natural magnet—its ability to attract pieces of iron and its directive tendency when freely suspended—were known to the Chinese and the Greeks at a very early date. The German naturalist Alexander von Humboldt, indeed, would have had us believe that the Chinese knew of the magnetic compass in the year 1100 B.C., though this view is now considered erroneous, having been based on a misinterpretation of Chinese records. It has, however, been confirmed that the magnetic compass was described by a Chinese writer around A.D. 1080, sometime before it was specifically referred to in Western Europe. Certain scholars have interpreted passages in Homer's *Odyssey* as indicating that the loadstone was known to Greek sailors at the time of the siege of Troy.

While our information about the extent of knowledge of magnetism in antiquity is uncertain, we can say with reasonable assurance that up to the middle of the thirteenth century all that was definitely known about magnetism was represented by the two properties of the loadstone familiar to the Chinese and Greeks. Many authors paid their respects to the phenomenon in exaggerated, mystical fashion. At various times during the Middle

Ages magnets were asserted to be a cure for all sorts of diseases, from tooth-ache and headache to gout and dropsy. One author maintained that magnets were of great value in disputes—possibly the basis of another author's belief that they could be used to reconcile husbands to their wives. (Precise details of the method of use in this connection are lacking.) It was believed that a magnet that had lost its power of attracting iron would regain its power if it were immersed in the blood of a buck or a goat, and some supposed that a magnet rubbed with garlic would lose its magnetic property. Stories were also told of the famous magnetic rocks in the Indian Ocean, which were said to draw the nails out of the ships that came near them.

Judged by modern standards, the first serious treatise on magnetism is generally believed to be that of Petrus Peregrinus, the Picard philosopher who, in 1269, wrote a letter to a soldier friend describing experiments he had performed with loadstones. Paul Fleury Mottelay calls this letter the earliest known treatise on experimental science.[8] Though this is a rather strong statement, Peregrinus' letter was a remarkable document. The letter indicates that Peregrinus was familiar with the poles of the loadstone and the fact that unlike poles tend to attract while like poles tend to repel one another. In it he also tells how a piece of ordinary iron may become a mag-net on being rubbed by a loadstone, evidently foreshadowing the produc-tion of artificial magnets. Peregrinus was the first to suggest studying a loadstone in the form of a sphere—a "terrella" (little earth), as it came to be called. Whether this should be taken to mean that he realized the earth was a magnet is a moot question. At any rate he made an important ad-vance in the use of the magnet as a compass by giving explicit directions for the construction of two types of compasses. It also appears that he ob-served the variation in the compass needle—that it does not in general point to the true geographic north. From the standpoint of a modern physicist, the only serious blot on the scutcheon of this ingenious man was his suggestion that a perpetual-motion machine might be devised using a loadstone. However, his device does not appear materially more implausi-ble than most of those that have been proposed by indomitable inventors of recent times. Peregrinus, who must have been a remarkable man, knew much more than his thirteenth-century contemporaries about the nature of magnetism. One theory is that he owed much of his knowledge to his distinguished contemporary Roger Bacon, who indeed refers to him in very flattering terms. It seems certain that many of the ideas of William Gilbert were suggested by the work of Peregrinus.

The period from Peregrinus' time to the middle of the sixteenth century witnessed the widespread development of navigation and the increasing use of the magnetic compass, with, however, only a very gradual increase in the understanding of its vagaries. Columbus, of course, was aware of the variation of the compass, or declination, as it is now called (the angle between true north and the direction of the compass needle), and more important still, the change in declination with longitude. On his initial voyage to the west he passed the agonic line, or line of zero declination, which caused some consternation when first observed by his navigators. By the year 1550 the dip or inclination of the magnetic needle (the angle between the horizontal and a compass needle allowed to swing freely in a meridian plane about a horizontal axis) had been observed by a German named Georg Hartmann. (The discovery is commonly attributed to the English compass manufacturer, Robert Norman, who is the authority for the statement that the dip at London in 1576 was 71° 50'.)

From the fourteenth and fifteenth centuries little else of importance has come down to us about fundamental magnetic phenomena. Paracelsus (1493–1541) collected assiduously all the hoary tales of the wonderful properties of the loadstone and added a few new ones, indicating that he was a better storyteller than experimenter. An Italian named Giambattista della Porta (1538?–1615) took the trouble to examine the old fables and in some cases performed experiments to demonstrate their objective falsity. Beyond this he does not appear to have added fundamentally to the knowledge of the subject.

The great systematizer of magnetic philosophy and science was William Gilbert (1540–1603), whose period of activity was the latter half of the sixteenth century and whose great book *De magnete,* published in 1600, was the fruit of seventeen years of labor and research.[9] During the time when Gilbert was carrying on his investigations, the Renaissance had been flourishing for more than a century, and science, no less than literature, was feeling the effects of the new enthusiasm for learning. It is true that much of the scientific side of the intellectual rebirth was a rediscovery of, and slavish commentary on, the scientific speculation of the Greeks and of Aristotle in particular. Nevertheless, the age produced a number of individuals who showed a decided tendency to strike out for themselves and to find out how things really behave, independently of what others had said about them. Among these were the versatile genius Leonardo da Vinci and the path-breaking astronomer Copernicus. While Gilbert was experimenting, his countryman Francis Bacon was urging experimentation as the only real way of adding to human knowledge. On the Continent Tycho Brahe was making the careful observations of planetary motions that were later to be so useful to Kepler in establishing the orbits of the bodies of the solar system. Mechanics was on the verge of a new method of approach with the work of Galileo, which marked the birth of what we may justly call the method of modern physics. It was a time when timid souls could satisfy their thirst for learning by drinking deep at the fount of the ancients but when there was also a spirit in the air that encouraged others to look to Nature herself for their knowledge.

Gilbert was one of the latter. In the Preface to *De magnete* he emphasizes that true philosophers are those "who not only in books but in things themselves look for knowledge," for "in the discovery of secret things and in the investigation of hidden causes, stronger reasons are obtained from sure experiments and demonstrated arguments than from probable conjectures and the opinions of philosophical speculators of the common sort."[10] Gilbert followed Bacon's advice: he not only talked about the desirability of performing experiments but actually carried them out and so laid a secure foundation for the future study of magnetism and of electricity as well.

Born in England in 1540 at Colchester, in Essex, Gilbert—or as the family apparently preferred to spell the name, Gilberd—attended the grammar school of his native town and entered St. John's College, Cambridge, in his eighteenth year. The records show that he was awarded the bachelor's degree in 1560, the master's in 1564, and the degree of doctor of medicine in 1569. In his travels thereafter on the Continent he acquired another doctor's degree in medicine and carried on a medical practice. On his return to England in 1573, he continued as a practicing physician in London and achieved a high place in the profession, becoming, in 1600, president of the Royal College of Physicians. Gilbert's prestige became so great that Queen Elizabeth appointed him one of her physicians-in-ordinary and settled a pension on him. Precisely when this appointment was made is not known, but Gilbert kept the position until the death of the Queen on March 24, 1603. He was continued in the same post by James I but died on November 30, 1603. (This was a year in which the plague flourished, though it is not known whether it was the cause of his death.) Gilbert was buried in Colchester, where a monument to him was erected by his brothers. Gilbert's library and scientific apparatus were bequeathed to the College of Physicians but were lost in the great fire of 1666. Only one known portrait has survived. Thomas Fuller described Gilbert in the following words: "One saith of him that he was Stoicall, but not Cynicall, which I understand Reserved, but not Morose, never married, purposely to be more beneficial to his brethren. Such his Loyalty to the Queen that, as if unwilling to survive, he dyed in the same year with her, 1603. His Stature was Tall, Complexion Chearful, an Happiness not ordinary in so hard a student and retired a person."[11]

Though few details of Gilbert's life have survived, we can form some opinion of the way in which he thought about science by examining his writings and, in particular, the famous *De magnete*. His book may well be considered the first extensive treatise on magnetism. In the opinion of many authorities it also marks Gilbert as the foremost scientist of Elizabethan England. However, it does not appear that he was actually so regarded in the England of his day, though his book created a considerable stir on the Continent and made a decidedly favorable impression on his contemporary, Galileo, then engaged in his epoch-making work on the foundations of mechanics. In his *Dialogues concerning the Two Great Systems of the World* (first published in 1632), the Italian scientist makes one of his characters speak of Gilbert in somewhat exaggerated terms:

I have the highest praise, admiration, and envy for this author, who framed such a stupendous concept regarding an object which innumerable men of splendid intellect had handled without paying any attention to it. He seems to me worthy of great acclaim also for the many new and sound observations which he made, to the shame of the many foolish and mendacious authors who write not just what they know, but also all the vulgar foolishness they hear, without trying to verify it by experiment; perhaps they do this in order not to diminish the size of their books.

There is, to be sure, one critical note in this encomium:

What I might have wished for in Gilbert would be a little more of the mathematician, and especially a thorough grounding in geometry, a discipline which would have rendered him less rash about accepting as rigorous proofs those reasons which he puts forward as *verae causae* for the correct conclusions he himself had observed.[12]

It might be supposed that Bacon, whose philosophic dictates Gilbert was largely following in practice, would have welcomed the new treatise with enthusiasm. He did indeed approve of Gilbert's experimentation but viewed with decided disfavor Gilbert's attempt to theorize extensively on

the basis of his experimental results.[13] Whether Bacon's adverse criticism
had a serious effect on the success of Gilbert's book is an open question. It
is certain, at any rate, that relatively little attention was paid to it in England until the early nineteenth century, when Gilbert and his work were
rediscovered and his early use of the scientific method was praised somewhat excessively.[14]

The book itself is a substantial treatise, discursive in style, which, in the
English translation of Mottelay, extends to more than 350 octavo pages. It
is profusely illustrated with line drawings to clarify the description of the
experiments and arguments of the author. The six books into which it is
divided are organized more or less coherently with respect to the topics
treated. Book I opens with a historical review of the previous discoveries in
magnetism and a frank exposure of myths and errors, continues with a fac-
tual account of the chemical and physical nature of the loadstone (with a
clear statement of its relation to iron ore), and ends with the revelation of
Gilbert's theory that the earth is a huge magnet. Book II discusses attraction, and after introducing at the beginning the celebrated comparison
between magnetic and electric attraction, is devoted to a thorough discussion of the attractive properties of magnets. Book III takes up the directive
force exerted by the earth on a loadstone or an artificial magnet made from
a loadstone. Book IV discusses the declination of the magnetic compass.
Book V discusses the dip or inclination. Finally, Book VI treats exclusively
of the earth as a spherical magnet and gives the author an opportunity to
express his adherence to the Copernican view of the rotation of the earth
on its axis. There is considerable unsupported speculation in Book VI, and
the connection between the earth's magnetic properties and its rotation,
though emphasized as significant, is nowhere clearly explained.

According to the Preface to *De magnete,* its author's main reason for
carrying out his investigations and writing his treatise was to put forth and
support a new theory of the nature of the earth. This theory is probably
less interesting to the modern student of magnetism than to the student of
Renaissance philosophy. Much more important for modern physics is the
fact that Gilbert did not content himself with mere verbal philosophizing
but carried out actual experiments and made an honest attempt to found
his theories on these experiments. In his experiments, furthermore, Gilbert
observed things apparently overlooked by his predecessors. Previous philosophers had found that a magnet possesses polarity and that while opposite poles attract, like poles repel. It had also been observed that arti-

ficial magnets can be made by rubbing iron on a natural magnet. But Gilbert went further and discovered that iron can be magnetized without touching a magnet directly. In other words, he discovered magnetic induction through space. Since most magnetization in modern instruments takes place in this way, perhaps it deserves to be ranked as Gilbert's most significant discovery. Thus, when a heated bar of iron is placed along the meridian and allowed to cool in this position, it is found to be magnetized and its north pole to be pointing to the north. Gilbert realized—and this also was probably his own discovery—that heating a magnet to red heat destroys its magnetism. He learned as well that if the demagnetized iron or iron ore is allowed to cool when lying in the proper direction or near a strong magnet, it will become remagnetized. The effect of heat on magnetism has been the starting point of much modern research on the subject, and in this, too, Gilbert was a pioneer.

To describe magnetic experiments more clearly Gilbert invented the concept of the magnetic field. An experiment in which he evidently took great pleasure involved placing very short pieces of iron wire on a loadstone in the shape of a sphere (a terrella). If the loadstone is sufficiently strong, the short wires stand up from the surface at all points except on the magnetic equator, where they lie tangential to the surface. At the poles the wires stand perpendicular to the surface, while at intermediate points they are oblique to the radii drawn to these points. Gilbert used the orientation of the wires as an indication of the direction of the magnetic force in the region surrounding the loadstone. This region he called the *orbis virtutis* ("region of strength" or, as we now call it, the magnetic "field of force"; it is also referred to as a sphere of influence within which the magnet can act on other pieces of iron). Since Gilbert probably had little or no conception of force from the dynamic point of view, his use of the term must be regarded as merely qualitatively descriptive. There is no evidence that he connected the decrease of the magnetic influence with distance from the magnet with any geometric property: he evidently had no glimpse of the inverse-square law. This is scarcely surprising, for the Newtonian epoch was still nearly a century away. On the other hand, we may justly share Galileo's regret that Gilbert refrained from a quantitative study of the magnetic field. While Gilbert quoted with approval a scheme of Porta's to estimate the strength of a loadstone by means of a balance, he presumably did not care to follow Porta's scheme, which has distinct quantitative possibilities. Mechanics needed the spark supplied by the genius of Galileo,

Huygens, and Newton before it could be applied successfully to magnetic phenomena. Some modern commentators have thought that Gilbert had a notion of the concept of mass as distinct from weight.[15] And some translators of *De magnete* have used the word mass in the discussion in which the strength of a loadstone is said to be proportional to its size—assuming, of course, that the various pieces under comparison are cut from the same stone. However, a careful reading does not show that Gilbert had any notion of mass in the dynamic sense; his idea hardly went beyond the concept of quantity as measured by size. This is another illustration of the great difficulty encountered by the modern scientist in interpreting the work of his early predecessors.

Without doubt, one of the reasons why Gilbert's book impresses us today as a monumental advance in the science of magnetism is that he devised new and greatly improved apparatus for his experimental investigations. The fact that he did so is closely connected with the growth during his age of the urge toward experimentation as the ideal procedure in the description of nature. Experiments demand apparatus, and it is clear that the success of scientific research rests on the development of appropriate devices. These devices need not be elaborate; some of the greatest discoveries in physics have been made with what today seem to be very simple apparatus. The point is that the apparatus must be appropriate for the subject under study. This is where the genius of the successful investigator displays itself. Gilbert introduced and used widely a very simple instrument, the "versorium," a light compass needle mounted on a sharp point so as to rotate freely on a vertical axis. Previous investigators had studied the behavior of magnets by floating them on water by means of wood or cork. But Gilbert's versorium, simple as it was, proved to be a much more flexible instrument and better adapted to a larger field of investigation. With it he was able to repeat under better-controlled conditions all the old experiments of Peregrinus, Porta, and others—among them the experiments on the law of polarity, the appearance of new poles when a magnet is broken in two, and the destruction of magnetism with heat. He was also able to study the new effects of induction that were his own significant discovery.

The versorium played a conspicuous role in what was probably Gilbert's greatest contribution to science, his recognition that the earth is a magnet. It is impossible to read his book without being aware of the delight he took in the analogy between the behavior of the little versorium on the surface of the terrella and the behavior of the compass, or dipping needle, on the surface of the earth. Whereas his predecessors had commonly attributed the directive tendency of the compass needle to some extraterrestrial cause —some "property of the heavens" like the action of the pole star—Gilbert saw clearly from his experiments on spherical loadstones that this attribution was wholly unnecessary: the simple assumption that the earth is a huge magnet is sufficient to account for all the phenomena observed by mariners and others using the magnetic compass. This is a prime example of Occam's principle of parsimony.

Gilbert devoted four of the six books of his treatise to the earth as a magnet. Much of his attention was necessarily focused on the important properties of declination and inclination, and he suggested new designs for constructing mariners' compasses and dipping needles. Gilbert's design of the former was not efficient and, though widely used during the eighteenth century, was discarded in the nineteenth. On the other hand, the dipping needle in use today is much like the one he designed. Since complete knowledge of terrestrial magnetism is not possible except through careful observations at many places on the earth's surface, it is not surprising that Gilbert's views on the subject have undergone considerable revision since his time. He was wrong, for example, in stating that the declination is constant at any one place on the earth. Evidently the previous observations at his disposal were not sufficiently accurate to show the small but definite secular change. It is interesting to note that this effect was actually discovered by Henry Gellibrand of Gresham College, London, who pointed out, in 1633, that the declination in London had changed from 11° E. in 1580 to 4° E. in 1633. The declination there became 0° in 1659 and then shifted west, reaching the maximum value of 24.5° W. in 1823. It has decreased again since that time, and it is estimated that it will be 0° again in 2139, an interval of 480 years from the last time the declination was 0°. If indeed the phenomenon is truly periodic, a result that we shall have to leave to our descendants to settle, the complete period would appear to be of the order of 960 years.

The observed variation in declination from place to place was a challenge to Gilbert's ingenuity. He proposed the interesting theory that declination is due to irregularities in—specifically, to elevations on—the earth's surface. He sought the reason for the easterly declination observed at his time along the western coast of Europe in the greater attraction of the mountainous continental land to the east compared with the lesser attraction of the ocean to the west. As an interesting generalization of this theory Gilbert argued for the existence of an open water passage in northern Europe from the Atlantic to the Pacific, for in his view the large westerly declination prevailing at that time in Novaya Zemlya indicated no continental land extending toward the east in the arctic regions. Gilbert therefore favored a northeast over a northwest passage. His whole argument founders, of course, on the actually existent secular variation of the magnetic properties.

Porta, and before him Peter Plancius, had thought that the declination varies in a regular way with the longitude and could provide a method for estimating longitude independently of astronomical considerations. Gilbert took pleasure in exploding this theory with ample experimental evidence to the contrary. Then, interestingly enough, he fell into a somewhat similar error himself with respect to the magnetic measurement of latitude when his close investigation of the behavior of the versorium on the magnetic terrella led him to formulate a very precise relation between the dip and the latitude. It is one of the few bits of genuine mathematical theorizing in *De magnete*, but unfortunately the reasoning is by no means entirely clear and would scarcely have satisfied Galileo. Gilbert tried to apply his formulation directly to the earth, undoubtedly believing that he would revolutionize the art of sailing and render mariners independent of the stars. The actual magnetic lines of equal dip (isoclinic lines), however, do not follow at all precisely the parallels of latitude, as any good magnetic map shows, so Gilbert's fine theory had to go the way of Porta's.

From the standpoint of modern physics, Gilbert was justified in believing the variation of dip with latitude to be a much more definitely predictable effect than the variation of declination with longitude. On the ideally magnetized terrella there are only two definite poles, or points at which the dip is 90°, and because of this symmetry there can be no declination at all if the magnetic axis is treated as the geometrical axis of the terrella. On the other hand, there is a definite clear-cut variation of dip with latitude on the terrella, which led Gilbert to the mistaken conclusion that the observed declination variation on the earth is due to local irregularities and therefore of no use in predicting position; actually, the observed dip ought to bear a close relation to the latitude. He should have realized, of course, that the same irregularities might also play havoc with his theoretically computed dip-latitude relation.

Gilbert's dip-latitude relation is an interesting puzzle in the history of physics. Because the relation was not expressed analytically but was thrown into geometrical form and was not explained very clearly, it is up to the historian of science to determine the basis for what Gilbert did.[16] Gilbert prepared an elaborate diagram from which, by following his instructions, the dip for any latitude can be read, and in 1602 an Englishman named Thomas Blundeville published, in an appendix to a book called *The Théoriques of the Seven Planets,* a table of numerical values of the dip-latitude relation based on Gilbert's diagram. (This was intended for the use of navigators, but it is not known how widely it was employed.) The present author has succeeded in finding a trigonometrical formula that gives results in precise agreement with Gilbert's diagram. It is likely that Gilbert was skilled in the use of trigonometry, but it is not clear that he used it in his construction, which seems to have been developed purely through intuition based on his experiments with the versorium on the terrella.[17] If Gilbert's dip-latitude relation is compared with that for a uniformly magnetized sphere as worked out analytically by Karl Friedrich Gauss in 1839, there is no wide difference, particularly at latitudes below 30°. Unfortunately the earth is not a uniformly magnetized sphere.

Gilbert believed that the earth is a great magnet as we do today. Gilbert, however, believed that this was so because the earth is somehow animate and possesses a soul. He could not see why cockroaches and worms should be endowed with the vital spark if it were lacking in our mother earth. Gilbert left the matter there and wisely did not try to explain the difference between a magnetic soul and a nonmagnetic soul. Since his time many people have tried, without complete success, to explain why the earth is a magnet. We know decidedly more about magnetism in general and about its relations with electricity in particular than Gilbert did, but the definitive solution of the theory of terrestrial magnetism has still to be made.

The last book of *De magnete* is curious and fascinating. In it the author abandoned the experimental method that he had pursued earlier with such success and launched forth into an almost dithyrambic appeal for the Copernican theory of the rotation of the earth based on its magnetic nature. The reasoning in this book, so utterly at variance with that of the earlier parts of *De magnete,* makes it difficult to understand why this last part was included at all. Possibly Gilbert wished to put himself on record as a convinced believer in the rotation of the earth and considered that since his book was essentially a treatise on the earth as a magnet, it was appropriate to include the argument on the Copernican theory. He may also have believed that the magnetic properties of the earth really do have an important bearing on its rotation. His reasons for believing this are by no means impressive, and the whole argument based on terrestrial magnetism is of a decidedly speculative character quite similar to the vague theorizings that Gilbert attacked with so much gusto in the works of other, earlier authors. He was human and subject to the frailties of us all, and it was probably extremely difficult for him to escape from the speculative tendency of the age, in spite of his zeal for experimentation. We can scarcely blame him for this, and it must be admitted that some of his arguments in favor of the rotation of the earth are plausible enough in themselves, although they have nothing to do with magnetism.

Since this excursion is primarily concerned with Gilbert's contribution to magnetism, nothing has been said about his experiments in electricity, which are fully described in Chapter Two, Book II, of *De magnete*. Here he summarized all the facts known up to his time on electrification by contact and very clearly contrasted magnetic and electric phenomena. The versorium invented for the study of magnetism turned out to be very useful also for experiments on electrically charged bodies. Gilbert's fame, however, rests principally on his magnetic experiments and theory. His work provided a great impetus to physical experimentation along all lines and set a fashion in the precise description of nature that has remained dominant ever since his day.

This excursion can scarcely be concluded more appropriately than by completing the quotation from Thomas Fuller that appears earlier in this chapter: "Mahomet's Tombe at Mecha is said strangely to hang up, attracted by some invisible Loadstone, but the memory of the Doctor will never fall to the ground, which his incomparable Book *De magnete* will support to Eternity."[18]

Pierre Gassendi and the Revival of Atomism in the Renaissance

Today atoms are taken for granted. The value of the atomic concept for modern physics is so great that scientists no longer are exercised over the question of the reality of atoms. But this attitude of detachment has not characterized every stage in the evolution of science, and there were epochs when it meant a great deal whether one believed in atoms or not and, more particularly, whether one stated this belief in public. This was peculiarly the case in the late Renaissance, the period to which the French cleric Pierre Gassendi belongs.

Atomism as a philosophy is very old. The Greek philosophic school represented in the fifth century by the rather shadowy figures of Leucippus and his pupil Democritus is best known to us by the adoption of their atomic theory into the philosophy of Epicurus (341–270 B.C.). The best summary of the atomic point of view in antiquity is provided in the poem of Lucretius (*ca.* 99–*ca.* 55 B.C.), *De rerum natura.* Unfortunately for atomism, Aristotle, who wrote the first serious textbook on physics, did not believe in atoms. He realized quite correctly that belief in their existence involves the notion of empty space or perfect vacuum, and he was quite sure that there is no such thing as a vacuum. Aristotle taught the continuum point of view, the theory that matter has a completely continuous constitution and that any so-called piece of material is indefinitely subdivisible.

The two fundamental viewpoints—the atomic and the continuum—had varying fortunes through the centuries until the Middle Ages, when the continuum theory triumphed and the atomic theory went into hiding. The philosophy of Plato and Aristotle found favor with the Church Fathers. When the latter considered atomism at all, it was to stress what they felt to be its essential atheism. Epicurus had taught that in the beginning there existed merely atoms and a void and that the universe had originated in the chance concretion of atoms flying around in quite random fashion; there seemed to be little room for God and the creation in this picture, and excoriation of Epicurus and atomism followed. Atomism in many theological circles became heresy.

These conditions, under which atomic theory made little headway during the Middle Ages, may have had a retarding influence on the advancement of science. Certainly the emphasis placed by the early atomists on motion as the most fundamental of physical phenomena should have led to an earlier revision of the Aristotelian theory of motion than was actually the case. The notion of atomism began to stir again in the early Renaissance. In the fifteenth century the German cardinal Nicholas Cusanus resurrected the term in his philosophical writings, but this apparently had little influence on his contemporaries and successors. In the sixteenth century an Italian, Giordano Bruno, introduced the atomic notion in a semi-mystical mixture of mathematics and physics but never applied it to the description of physical phenomena. Francis Bacon (1561–1626) revived the atomic theory of the Greeks, only to disagree with it. In Germany, in 1619, the physician Daniel Sennert introduced a corpuscular theory to explain chemical processes, and the Italian Sebastian Basso, another physician, who practiced in Paris, attacked Aristotle in the work on natural philosophy he published in 1621. (Significantly, publication took place in Geneva, not Paris.)

Basso's work was undoubtedly largely responsible for an interesting attempt to bring the atomic point of view to the attention of the citizens of the French capital. Three bold spirits—Jean Bitault, Antoine Villon, and Etienne de Claves—announced that they proposed on August 24 and 25, 1624, to discuss publicly, with experimental demonstrations, certain theses against Aristotle. (Their fourteenth thesis not only came out openly in favor of the atomic theory of Democritus and Epicurus but castigated Aristotle with some severity.) About one thousand persons came to see the show, but the performance failed at the last moment to come off, for the authorities, evidently sensing the possibility of a riot, forbade it and arrested De Claves. Villon, the soldier-philosopher, saved himself only by flight. (Nothing is known about what happened to Bitault.) The three would-be perpetrators of the crime against Aristotle were exiled from Paris, and it was made a crime punishable by death to publish such theses as theirs, or to hold such disputations, anywhere in the kingdom without the consent of the theological faculty of the University of Paris.

It was necessary for believers in atomism in the early seventeenth century to proceed with some caution. Even Descartes, though he introduced a corpuscular hypothesis into his principles of physics, did not go all the way toward the ancient atomistic theory of Epicurus, since he did not admit the idea of vacuum into his theory. In fact, just as Descartes' famous treatise *Le monde* was on the point of publication in 1634, the news came to him of Galileo's persecution by the Inquisition and of Galileo's abjuration. Descartes decided not to publish and kept to his decision until the powerful influence of Cardinal Richelieu was exercised in his behalf; only then did it become safe to proceed.

In the light of the preceding paragraphs it is surprising to find that Pierre Gassendi—the chief reviver of Greek atomism in the late Renaissance, the man whose work is chiefly responsible for the inspiration of Boyle and through him of Newton and hence the whole Newtonian school —was a devout priest of the Roman Catholic Church. Gassendi (1592–1655) is by no means one of the giants of physical science; his contributions to physics fade into insignificance compared with those of Galileo, Huygens, and Newton. But his stubborn determination to breathe new life into the doctrines of Epicurus at a time when it was unfashionable, if not positively dangerous, to do so justifies attention to his career and his ideas.

Gassendi's life was so completely devoted to religion and scholarship that even the most ardent biographer can find little excitement in it. Confronted by a situation of this sort a biographer is apt either to reconstruct the historical environment and place his hero in it or to immerse himself in the writings of the man and build a picture of him from what he had to say. The second course is not made any easier by the fact that Gassendi was a prolific writer, whose six large tomes of medieval Latin form a formidable barrier to the average physicist. No complete translation of his works into any other language has been made, though an abridged version in French was brought out by his pupil François Bernier in 1678.

Gassendi was born of peasant stock in the village of Champtercier in Provence, some four miles from the town of Digne. He was a child prodigy and at the age of four preached sermons from his window to the astonished edification of the passers-by. Sent by his parents to the college at Digne when hardly out of his swaddling clothes, by 1609 he had exhausted the well of knowledge there and moved on to Aix, where he studied philosophy. (His teacher is supposed to have remarked, "I don't know whether he is my pupil or my master.") At the age of twenty Gassendi became principal of the college at Digne. When he was twenty-two years old, he became a doctor at Avignon, and two years later he entered the priesthood on the advice of influential friends, who may have felt that he could take better advantage of his intellectual talents by doing so. When the chairs of theology and of philosophy at the University of Aix fell vacant, Gassendi won both chairs by the disputation method but turned over the chair of theology to his old teacher, retaining the chair of philosophy for himself. His choice reflected to a great extent his future interests and attainments: Gassendi apparently looked upon the Church as a means of livelihood, but his real interest was philosophy, which of course in his time meant science.

In the winter of 1624/25 the still-youthful philosopher went to Paris, where his reputation had preceded him and gained him the esteem of Descartes and the notable circle of savants who were the precursors of the famous Academy of Sciences. In 1624 Gassendi had already published his first book attacking Aristotle, *Exercitationes paradoxicae adversus Aristoteleos.* After the experience of Bitault, Villon, and De Claves, it would seem to have required considerable courage to come out against Aristotle, but all through his life Gassendi appears to have coasted along just on the edge of heresy. Perhaps in the little town of Digne things could be said about Aristotle that the venerable faculty of the University of Paris would not tolerate in the metropolis. At any rate, by 1625 Gassendi had adopted the Copernican theory and had written to Galileo congratulating him on his defense of it. Obviously the condemnation of Galileo by the Inquisition in 1633 came as a considerable shock. Gassendi weathered the storm by going about his business, evidently thinking his own thoughts but keeping his mouth shut.

Gassendi's career in the church involved much ecclesiastical politics and even litigation. For example, his provostship at Digne was secured only after a law suit that lasted seven years. (He usually won in such encounters. His biographers lay great stress on what a modest man he was, but he evidently got pretty much what he wanted.) In 1645—through the influence of Louis-Alphonse du Plessis de Richelieu, Cardinal-Archbishop of Lyon, Grand Almoner of France, and brother of the more famous Armand— Gassendi was appointed Royal Professor of Mathematics at the Royal College at Paris, now the Collège de France. Here he actively joined the group of philosophers who met more or less regularly in the cell of the Minorite Marin Mersenne. It was a distinguished circle that included the mathematicians Fermat, Désargues, and Roberval, as well as Pascal. In this setting Gassendi indulged his interest in experimental science. That he was no mean experimenter is shown by his observation of the transit of Mercury across the sun's disk in 1631; according to some authorities he was the first definitely to observe it. Gassendi was also interested in acoustics and made measurements of the velocity of sound in air as early as 1635. The value that he found was too high, as Mersenne showed by a more accurate determination; however, Gassendi did establish one important point, that the velocity of sound in air is independent of the frequency—a fact of which Aristotle was ignorant. Of course Gassendi made mistakes, too, believing, for instance, that wind has no effect on the velocity of sound.

In 1647 Gassendi published a life of the Greek philosopher Epicurus. For many years Gassendi had been leaning toward the atomic doctrine and had yearned to rehabilitate the tarnished reputation of the ancient atomist. Gassendi had long hesitated to come out in the open, however, and for the most part his views had been expressed in letters to his contemporaries, particularly Mersenne. He may finally have secured ecclesiastical approval for the publication of the life and philosophy of the hated and despised atomist because the ecclesiastic authorities were satisfied that he had succeeded in adapting the Epicurean atomic doctrine to orthodox Christianity. Certainly Gassendi always showed himself a very cautious writer; he was very different in this respect from Galileo, who liked nothing better than to poke fun at his adversaries. Voltaire later lampooned the theologian-philosopher as "the unsettled, uncertain Gassendi."

In 1648, at the peak of his fame, Gassendi's health broke down, and he returned to his ecclesiastical position at Digne. While there he received a pressing invitation from Queen Christina of Sweden to come to Stockholm to take the place at her court of Descartes, who had just died there. Gassendi, who may have heard of the troubles of his distinguished contemporary in trying to teach the boisterous queen the elements of natural philosophy, politely declined. In 1653 Gassendi was back in Paris, this time never to return to his old home. During 1654 his physical condition grew steadily worse, and on October 24, 1655, he died at the age of sixty-two.

The most elaborate, and presumably the definitive, exposition of Gassendi's ideas is to be found in his *Opera omnia,* published in six volumes in 1658 at Lyon and reprinted at Florence in 1727. The first volume contains most of his physical philosophy. It is scarcely necessary to say that it is not at all easy reading. Such common Latin words as *qualitas, facultas, atomus, calor, frigor, plenum, solidum, velocitas, levitas, gravitas, densitas,* and *raritas* appear frequently, but their meaning, in Gassendi's context, is uncertain. Every examination of such a work must again remind the physicist of the tremendous difficulties confronting the historian of science, whose task it is to penetrate behind the abstract language of other ages to the operational meaning and then to translate the latter into the equivalent scientific terminology of today. This problem is not so difficult in the case of an author who describes actual experiments and gives diagrams or pictures illustrating what he did and what he found. Galileo, for example, is open to little or no misinterpretation in much of his writing. When he says that he allowed a ball to roll down an inclined plane and draws a picture of the plane, there is no doubt about what he was thinking. This is particularly the case when Galileo puts his results in mathematical form, for mathematics is a universal language. Unfortunately, with Pierre Gassendi the case is different. He evidently wrote for his own time and assumed with reason that his contemporaries would have no trouble understanding his language. Though in his exposition of atomic theory he referred frequently to ordinary human experiences and introduced some interesting, if at times far-fetched, analogies, he described almost no experiments, used no diagrams, and—oddly, since his astronomical writings show that he possessed an adequate grasp of geometry—presented practically no mathematical discussion.

Gassendi broke away from scholastic philosophy by postulating the existence of a real world of physical objects, independent of physical bodies, and existing in space and time. For Gassendi, space is real but incorporeal; it is unmeasurable, infinite, and eternal. It had been so, long before God created the world, and would exist forever, even if God should destroy the world. It is absolutely immovable. Wholly independent of God, it was never created; in fact it is uncreatable and indestructible: it simply is. Space, according to Gassendi, is the frame of reference in which all physical objects are imbedded and all physical phenomena are observed. The same attributes of necessity, incorporeality, and infinite extent are also given to time, the medium in which physical events take place. It is flowing, or moving, extension in distinction from static extension.

Physical events in space and time form an everchanging sequence or flux, but Gassendi could not believe that this was typical of real existence. He believed that there must be a fundamental unchangeable something at the basis of all our perceptions, the *principium materiale* or *prima materia* (prime, or first, matter). This he considered to be the stuff of which all perceived objects are made. Here Gassendi was reverting to the ancient Greek doctrine of "principles" and departing from the scholastic philosophy of substantial and accidental forms. (According to the latter, a jug is a jug because it has the substantial form of "jug-ness," while it is light or heavy because it has the accidental form of lightness or heaviness, and so on.)

In Gassendi's theory atoms are prime matter, but the notion of particle or corpuscle that was fairly common in his time must not be confused with Gassendi's atom. To him atoms are the final and smallest components of all physical bodies. They are not mathematical points but have finite extension, even if they are too small to be observed directly by the senses. They are so solid that no natural force can break them. Gassendi admitted that all actually observed objects can be subdivided into smaller pieces. This seems obvious, but he deliberately refused to postulate the possibility of the indefinite extension of this process. Mathematically it may be imagined, but physically it is impossible. To believe in its possibility is, according to Gassendi, to nullify the fundamental axiom of all scientific reasoning: from nothing comes nothing, and nothing cannot be converted into something. In our modern sophistication we may well assert that this is a futile argument, but perhaps it was a good thing for the development of physics that Gassendi took it seriously.

Gassendi assumed that all the substance in the world is located in the atoms, that everything else is empty space. In other words, he believed firmly in the existence of a vacuum. This is interesting, since during Gassendi's lifetime Evangelista Torricelli invented the barometer and made his discovery of atmospheric pressure and Pascal suggested the famous confirmation of this result by observing the fall of the barometer with an increase of altitude. Pascal apparently had no doubt that the space above the mercury surface in the barometer tube is a genuine vacuum; Gassendi, however, did not believe in the Torricellian vacuum. His vacuum was a hypothetical one assumed for the purpose of providing empty space for the atoms to move around in.

The most important property of atoms for Gassendi is their mobility or tendency to motion, which he called *gravitas*. Their size and shape are also significant, and these three properties must be sufficient to account for all the diversity of observed phenomena; the variety of possible sizes and shapes is very large but—contrary to the thought of the ancient atomists—not infinite. Gassendi wisely dodged infinity; he must have realized that too many clever people had burned their fingers with that tricky concept. "Large enough, but not indefinitely so," seems to have been his motto. Arguing for atoms of differing shapes, Gassendi pointed to the various kinds of crystals observed in nature and reasoned that it is natural to assume, for example, that the atoms making up common salt are cubical in shape because the salt crystal takes the cubical form. On the other hand, he held that certain atoms may be of very irregular shape and even have hooks on them for attaching themselves to other atoms to form groups with special properties—molecules, as these groups ultimately came to be called.

When God created the atoms (Gassendi is careful to give the Creator the credit for being the First Cause of things, though he usually disregards Him thereafter), He endowed them with an urge to motion that is the same for all atoms and that they can never lose. This mobility, Gassendi maintained, results in actual motion with a high velocity in a vacuum. (Precisely how high is not indicated.) In space occupied by other atoms there is continual bumping, and the velocity of the atoms may be retarded. Gassendi assumed, however, that while in motion the atom always has the same velocity. Its average velocity may indeed be less than the standard vacuum value because of rest periods due to collisions, but Gassendi did not appear to worry over the fact that this involves infinite deceleration and acceleration. (Unconsciously he was getting involved with infinity in spite of himself.) In any case, his concepts of motion are very vague, and he made no attempt to formulate them mathematically; if he had read Galileo's expositions on mechanics, he either did not understand them or did not choose to follow them. Here he missed a golden opportunity, for he might have created a full-fledged kinetic theory of matter if he had applied Galileo's principles of motion to his own theory about atoms and had worked out the consequences. This task was left for Robert Boyle to suggest and Daniel Bernoulli to carry out about a century later.

Gassendi's view of the observed properties, or qualities, of bodies was essentially empirical. He held that all knowledge of such things comes from our sense perceptions and that it is the task of the philosopher to describe these perceptions in terms of a postulated set of constructs. Such a set is provided by the atoms, which by their size, shape, number, arrangement, and motion are responsible for all observed phenomena. For example, one of the most obvious differences between substances is their density, and a block of lead of given volume is said to be more massive than a block of wood of the same volume—the density of lead exceeds that of wood. According to Gassendi, this is because in the wood there are relatively few atoms in the given space, or to put it another way, because there is more vacuum, or empty space, in the less dense substance. Because Gassendi thus reduced density to a numerical property and apparently paid no attention to the possible difference in mass of the constituent atoms, nineteenth-century physicists and chemists could scarcely avoid considering Gassendi's view absurd. But with the advent of nuclear physics and the assumption that all atoms are composed essentially of three fundamental building blocks—protons, neutrons, and electrons—scientists are now in a sense re-establishing Gassendi's numerical criterion.

To Gassendi, who was very thoroughgoing in his atomism, light, heat, cold, and sound are all atomic phenomena. According to him, a body is transparent because there are vacant spaces between the atoms for the light particles to go through; in opaque bodies these *vacua*, or pores, are present in smaller numbers or are absent altogether. Gassendi also believed that there are very small, round, and fast-moving atoms that can be emitted by certain substances, like flames. He believed that these atoms are able to penetrate the pores of other substances and push apart their constituent atoms and that this process constitutes heating and the concomitant change in size of the heated body. Not content with this not-too-unreasonable application of his theory, Gassendi explained the nature of cold by another set of atoms, much larger and slower than the heat atoms, which are tetrahedral and which, when they penetrate into bodies, have a tendency to unite the constituent atoms. Gassendi's explanation of cold, violating as it does the fundamental principle of economy in scientific thought, is less plausible than his theory of heat. From the modern point of view Gassendi worked with too many different types of atom.

Gassendi's interest in acoustics led him to several experimental observations on sound. He correctly explained the echo as resulting from the reflection of sound from surrounding surfaces, though this of course had been suggested in ancient times. In his thoroughgoing atomism he considered sound to be due to characteristic atoms emitted by the sound-producing body. He believed that when these atoms enter the ear, they react with a particular set of atoms in the ear to produce the sensation of hearing. He also believed that the velocity of sound is simply the speed of the sonic atoms, whereas the pitch is proportional to the number emitted per unit time, by analogy with the notion of frequency in the wave theory.

Any theory of matter and mechanics must account for the fall of objects toward the earth. This necessity gave Gassendi considerable trouble, for his fundamental philosophy forbade him to consider free fall as due to some inner quality of bodies that attracts them to the earth—it was too *ad hoc*. For Gassendi, the fall had somehow to be associated with action taking place directly in the atoms of the falling object. Since this action could not be action at a distance but had to be immediate and through direct contact, Gassendi was forced to assume that the earth emits streams of what might be called "gravitation particles"—particles so effectively continuous in character as to possess the rigidity of stretched strings and to act like tentacles to grasp the atoms of the falling body and pull them to earth. Since Gassendi assumed that these streams of particles emanate radially, their gripping power decreases with their distance from the earth. Just how these particles bring about the increase in velocity of fall (now known to be the acceleration due to gravity) is not made clear, for Gassendi lacked Galileo's pragmatism.

Gassendi's atomic theory is naïve; as a direct contributor to the advancement of modern atomic physics, the philosopher of Digne missed the boat. Still, he resurrected the atomism of ancient times, made it fashionable, and clothed it in the habiliments of ecclesiastical orthodoxy. By doing so, he made it possible for sincerely religious philosophers, like Boyle and Newton, to accept the atomic theory and apply it in their much more significant speculations. (Boyle, especially, frequently acknowledged Gassendi's influence and warmly welcomed Gassendi's attempt to reconcile religious teleology with Epicurean atomism.) Today this may appear a very trivial matter, but in the seventeenth century it was a very serious one. Many of the foremost natural philosophers were deeply religious men and were much concerned lest their scientific theories undermine their faith.

Galileo Galilei and the Motion of Falling Bodies

This excursion will be limited to a survey of some of the ideas about motion prevalent in Galileo's time and of the relation between them and his own concepts. The discussion will also illustrate the kinds of questions that the physicist who approaches the history of his subject raises.

In August, 1638, John Milton, then a young man of twenty-nine, visited Galileo, a blind old man of seventy-four, at a villa in Arcetri, a mile from Florence. "There it was," Milton recalled, "that I found and visited the famous Galileo grown old, a prisoner to the Inquisition, for thinking in Astronomy otherwise than the Franciscan and Dominican licencers thought."[19] Galileo's course was nearly run when Milton visited him. In that year Galileo's last great work, *Discourses and Mathematical Demonstrations concerning Two New Sciences pertaining to Mechanics and Local Motions,* was published in Leiden. Because of difficulties with the Inquisition the book could not be brought out in Italy, and when it finally did appear in his own country, the author had reached the stage where he could only fondle the volume and turn the pages, for loss of sight prevented him from following the printed words so important in the history of classical mechanics.

The progress of science, like other phases of human evolution, is full of ironies. Galileo's persecution by the Holy Office resulted primarily from his astronomical discoveries and his support of the Copernican theory. These contributions were noteworthy and have long excited the popular imagination, but it is now agreed that his greatest contribution to modern civilization was his theory of mechanics, which laid the groundwork for his distinguished successors Huygens and Newton. Although it is unlikely that many of his contemporaries understood fully the significance of his method of theorizing in mechanics, it was to make physics the most successful of all the sciences in the description of nature.

During the sixteenth and seventeenth centuries there was an evolution from medieval to modern thinking in science. The word evolution should be emphasized, for it has become fashionable to talk about the twentieth-century revolution in physics, and the layman may easily get the impression that physics progresses by the violent overthrow of accepted ideas. That this is far from true is well illustrated by the work of Galileo. Though Maxwell suggested that the theory of the magnetic field due to electric currents sprang full-grown from the brain of André Marie Ampère, whom he called the Newton of electricity, it is erroneous to assume that the laws of motion sprang in their final form from the head of Galileo. Galileo (1564–1642) was a child of his era, as are all scientists, and scientific thought is no more timeless than any other, for it is a function of the spatial and temporal environment. Because of the scientist's preoccupation with concrete experience and actual apparatus, the barriers of space and time are less apparent than they are in the arts or in letters. It must be remembered, however, that science is more than the performance of operations with gadgets, that it involves the description and understanding of experience—which is, of course, impossible without the introduction of abstract concepts. And the construction of concepts is an activity that is dependent on the scientist's physical and mental environment.

To the modern physicist motion seems to be so easily explained that it seems surprising there could have been so much disagreement about its nature in classical antiquity and the Middle Ages. How could Aristotle possibly have gone so wrong in his theory of motion? The facts now appear so simple that the elaborate verbalisms of the *Physica* are read only with the greatest impatience—until it is recalled that Aristotle was not interested in the kinds of questions now associated with motion. To the heirs of Galileo the theory of motion requires that the concepts and postulates lead by deduction to actually observed motions. But this was not at all what Aristotle meant by a theory of motion. Starting out with the postulate that the natural state of order on earth is rest, he was forced by his philosophy to seek a cause for terrestrial motion as a transient disturbance. To do this he divided all terrestrial motions into two classes—natural and violent. According to this philosophy, every object has a natural place in which it will remain at rest unless disturbed, and natural motion is the result of an attempt on the part of a body to regain the natural place from which it has been displaced. (The freely falling body is striving to get to its natural resting place.) On the other hand, every violent motion is a disturbance of order; its explanation gave Aristotle more trouble. He was committed to the idea that the cause of violent motion can be found only in direct material contact with the moving body and shunned action at a distance, as does present-day field physics; but Aristotle had no "field" to fall back on, as had Faraday and his nineteenth-century contemporaries. To Aristotle, violent motion demanded the continuous action of an external mover joined to the moving body. (According to his theory, when the external mover ceases to move, or becomes separated from the body, the latter also ceases to move.) The motion of a projected body—an example of continuous motion with no apparent mover—thus gave Aristotle considerable difficulty, and he was forced to explain it by some reaction of the surrounding medium. His explanation failed to satisfy all his scientific successors of the Middle Ages, and even his faithful adherents could not wholly swallow it; the independent thinkers made it the spearhead of their attack on Aristotelian physics in general.

Another peculiar feature of Aristotle's physics that caused trouble for his scientific descendants was his denial of the possibility of the existence of a vacuum, which was ingeniously related to his theory of motion. Since Aristotle held that the velocity of a freely falling body is proportional to the quotient of the weight of the body and the resistance of the medium, it necessarily follows that in a vacuum (where there is no resistance) all bodies will fall with infinite velocity. But such velocity has never been observed. Moreover, violent motion could never take place in a vacuum, since there would then be no medium available to serve as a mover. It is small wonder that Aristotle could not bring himself to believe in the real existence of a vacuum, although he might accept it as a mathematical fiction, the abstract construct for Euclidean space. According to Aristotle, only mathematical figures, not physical objects, may be placed in mathematical space; he was very reluctant to mingle physical reality with mathematical abstractions. (It is a curious fact that Archimedes, the greatest physicist of antiquity, apparently left no works on motion. From his treatment of the problems of static equilibrium, it might be expected that he would have anticipated Galileo if he had made a study of motion. Possibly the subject was not sufficiently amenable to mathematical treatment to suit his taste.)

Galileo's teacher at the University of Pisa was the Florentine Francisco Bonamico, whose lectures were published in 1611 in a large volume entitled *De motu*. Bonamico might have been expected to have exerted some influence on Galileo's thought, but Galileo was evidently an independent pupil, and if he received any stimulus from his teacher, it was to disagree with Aristotle and look elsewhere for an explanation of motion. Bonamico represented medieval Aristotelianism in an elaborate but rather confusing way. At any rate, it is extremely difficult for a modern physicist to think of motion as he did. Bonamico undoubtedly recognized the difficulties of Aristotle's theory of falling bodies. Outstanding among them is the puzzle as to why a constant cause (the weight of a body) should produce a variable effect (the accelerated motion). Aristotle held that since all freely falling bodies tend to reach their natural place, it is also natural that they should strive increasingly as they approach their goal. The medieval commentators evidently thought that this was an oversimplification. Wishing to connect the tendency to seek the natural place with some independently observed characteristic of the body, they selected the weight as the significant characteristic. Hence their dilemma. Inevitably, an Aristotelian like Bonamico was led to a solution by assuming that the resistance of the medium decreases during the fall or, at any rate, that its interaction with the motion somehow adds continually to the assumed constant effect of the weight.

There had been critics of Aristotle before Galileo, among them John Buridan and Nicole Oresme of the Parisian school of the latter part of the fourteenth century and, of course, Leonardo da Vinci. A critic of Aristotle closer to Galileo's period was Giovanni Battista Benedetti (1530–90). Galileo was probably familiar with Benedetti's *Diversarum speculationum mathematicarum et physicarum liber* (1585), though there is no explicit reference to him in Galileo's writings. In his treatise Benedetti summarized his adherence to the school of the "impetus" theory, already suggested by the Parisians. Benedetti's work disposed of Aristotle's theory of the effect of the medium on the motion of a body by insisting that every object, whether set in motion naturally or violently, receives an *impetus* (another Latin term, *impressio motus*, was also used), so that even when separated from the origin of its motion—for example, the hand—it can continue to move by itself. In natural motion the impetus increases without limit, whereas in violent motion its effect is to make the body thrown into the air lighter than normal. Moreover, when the motion is caused by hurling with a sling, more impetus can be stored and the resulting motion protracted. Benedetti was still following in Aristotle's footsteps in seeking a cause for violent motion, but he was searching for it quite differently and was impatient with Aristotle's errors. How medieval and yet, in another sense, how modern are the words with which he expresses his attitude toward the great philosopher of antiquity: "Such is the greatness and authority of Aristotle that it is difficult and dangerous to write against his teachings, and to me in particular since I have always held his wisdom a matter for admiration. Nevertheless, impelled by zeal for truth, by the love of which he himself, if he were living now, would be actuated, I have not hesitated in the interests of all to state wherein the unshakable foundations of mathematical philosophy force me to dissociate myself from him."[20]

It is clear from such writings as those of Bonamico and Benedetti that the problems of motion they discussed are quite different from the problems that interest physicists today. Benedetti, like Aristotle, is really intrigued by what are now called the initial conditions of the motion—in how it happens that when motion is communicated to a body by the hand, the body continues its motion after the removal of the hand. The modern physicist has no interest in this matter, which has been completely sidestepped by the introduction of the concept of inertia—an interesting illustration of one way in which problems are solved. Popular books and elementary treatises still expatiate on the errors of our forebears, who, it is forgotten, were often trying to solve problems that are no longer relevant. Physical explanation varies from age to age; there will never be an absolute criterion of its meaning.

Galileo's ideas on motion underwent a gradual development. This is very clear from an examination of his early work composed in Pisa and published under the title *De motu* in 1590. In this work the young philosopher appeared as a violent and not always discriminating anti-Aristotelian. In the enthusiasm of youth he was quite certain that Aristotle was completely mistaken. But, like Benedetti, Galileo was still fettered by the old Aristotelian tradition. He was still trying to discover why bodies move as they are observed to move and believed that the answer must be found in the impetus theory, the concept of impressed force. Not satisfied with previous work on the impetus theory, he gave it a thoroughgoing overhauling. The result makes strange reading for those accustomed to hear Galileo acclaimed as the founder of classical mechanics and modern physics.

The idea of impressed force probably was clearer in the work of Galileo than in that of any of his contemporaries. He visualized it as a quality that, like heat or cold, can be transferred from one body to another. When transferred, it becomes the property of the body to which it goes and no longer has anything to do with the body from which it has come. Aristotle's difficulty about the necessity of the presence of a continual mover attached to the moving body is thus avoided, but the impressed force always dies away eventually, since no motion endures forever. It is clear that the idea of inertia was far from Galileo's mind at this stage.

In *De motu* Galileo strenuously insisted that the velocity of a freely falling body is proportional to its weight. He recognized the initial acceleration of the falling body but was sure that ultimately the velocity would attain a constant value that could actually be observed if only the body could be dropped from a sufficiently high tower. This is strange language from the man who is popularly supposed to have revolutionized science by dropping balls of different weights from the Leaning Tower of Pisa to destroy the whole Aristotelian fallacy. (Indeed, though it is not certain that this experiment was ever performed by Galileo, Simon Stevinus of Bruges, in his *Oeuvres mathématiques*, published in 1634, states that in 1585 he had carried out the test and found that the difference in weight in freely falling bodies dropped at the same time from the same place appeared to have little, if any, influence on the time of fall.)

In his assumption of a constant final velocity for a falling body, Galileo was not far off, for it is now known that a constant velocity is actually brought about ultimately by the resistance of air. However, this is not another of those much-talked-about early anticipations of modern knowledge —the history of science is full of such pitfalls—for a careful reading provides no justification for such a conclusion. Galileo undoubtedly considered Aristotle's views on the influence of the medium through which the falling body travels so erroneous as to require an entirely new conception. According to Galileo, a ball rises when thrown up into the air by hand because the hand impresses sufficient "lightness" in it to overbalance the "heaviness" due to its weight; as long as the lightness exceeds the heaviness, the ball continues to rise. When the two become equal, the ball reaches its highest point, stops, and then starts to fall. However, the lightness has not entirely disappeared, and as long as any lightness remains it keeps the ball from attaining the final maximum velocity characteristic of its weight. This interesting point of view may not be completely original with Galileo (Hipparchus may have held it in antiquity), but Galileo went further. According to him, even a body dropped freely from rest from a high tower has lightness impressed in it, and this lightness provides the check that makes it begin slowly and pick up speed only as it falls. (Presumably the lightness results from the process of taking it up into the tower in the first place, but it is wise to resist the temptation of seeing in this the germ of the idea of potential energy.) Galileo claimed to have observed experimentally that, at the beginning of fall, light bodies fall faster than heavy bodies. This he explained by assuming that it is harder to impress lightness in a light body than in a heavy body. Hence at the beginning of free motion, light bodies have less lightness in them to oppose the downward motion.

As a vehicle for motion, Galileo would have nothing to do with the medium through which motion takes place, but he did not neglect the medium entirely. Like Benedetti, Galileo was greatly influenced by the work of Archimedes, and the concept of buoyancy, the central notion in Archimedes' principle of floating bodies, was the kernel of Galileo's theory of the impetus. Buoyancy was indeed the basis of his distinction between absolute weight and relative weight, and it led him to the conclusion that only in a vacuum do bodies fall with the velocity appropriate to their absolute weights. Galileo's numerous references to Archimedes are always laudatory; evidently Galileo felt that the ancient physicist was on the right track. It is significant that Galileo's first published work of scientific consequence, his little treatise on the hydrostatic balance (1586), was clearly in the Archimedean tradition.

The impetus theory probably received its most elaborate development from Galileo. If he had gone no further, he would have been recognized as a valiant critic of Aristotle and an eminent Renaissance philosopher. His ultimate success and fame, however, are the result of his realization of the essential failure of the impetus theory. Having tried to develop a better physics than Aristotle's and having failed, he examined the problem in an entirely different manner and discovered the law of falling bodies that he announced in 1604 in a famous letter to Paolo Sarpi. By this time Galileo had decided that it is more profitable to study precisely how bodies fall than to speculate on why motion takes place at all—to take motion as something given and examine it with care, rather than to try vainly to understand why human experience should include the phenomenon of motion. Galileo's decision was in a certain sense an expression of defeat, for it involved abstracting from the totality of experience something that seemed tractable to, and manageable by, the human mind. It was a momentous decision, for it implied a thoroughgoing break with the Aristotelian tradition and a determination to place the future development of physics on the Archimedean basis. It has colored the whole subsequent history of the science.

Galileo's new method appears clearly in his letter to Sarpi:

In reflecting on the problems of motion, for which I felt I lacked an absolutely sound and self-evident principle which could be used as an axiom in order to demonstrate logically the properties observed by me, I have finally arrived at a proposition which appears sufficiently natural and evident. Assuming this, I have been able to derive everything else, and in particular that the space traversed in natural motion varies directly as the square of the time . . . And the principle is this: that natural motion takes place in such a fashion that the velocity varies directly with the distance traversed . . .[21]

The important thing about this statement is not the precise character of the fundamental principle or axiom. It does not lead to the law of motion stated and therefore represents an error in mathematics—a pardonable error perhaps, since Galileo did not have the differential calculus at his disposal. The important issue is Galileo's abandonment of the purely empirical method of physical description by which each physical phenomenon is examined in turn and explained as closely as possible in terms of common sense. He now proposed to make physics a deductive science based on abstract concepts and purely postulational principles not susceptible to proof, hoping that on this foundation all observed phenomena might ultimately be logically deduced. (This is essentially the method of physics as it has been discussed in Chapter Two.)

Galileo's use of the method of modern physics is also clearly shown in the following passage in his *Two New Sciences*, in which, in his new approach, his original assumption of the dependence of the velocity on distance traveled is replaced by the more successful one of dependence on elapsed time:

At present it is the purpose of our Author merely to investigate and to demonstrate some of the properties of accelerated motion (whatever the cause of this acceleration may be)—meaning thereby a motion, such that the momentum of its velocity [*i momenti della sua velocita*] goes on increasing after departure from rest, in simple proportionality to the time, which is the same as saying that in equal time-intervals the body receives equal increments of velocity; and if we find the properties [of accelerated motion] which will be demonstrated later are realized in freely falling and accelerated bodies, we may conclude that the assumed definition includes such a motion of falling bodies and that their speed [*accelerazione*] goes on increasing as the time and the duration of the motion.[22]

Much effort has been expended investigating Galileo's reason for trying to base the theory of free motion on the hypothesis that the velocity of fall varies as the distance traversed. The simplest solution is to regard his attempt to do so as another remnant of the Aristotelian tradition that even the medieval critics of Aristotle were unable to discard. After all, "the farther from the origin of fall, the faster the fall," is one of the simplest of observations; it only remains to replace the qualitative connection with algebraic proportionality. To assume that the velocity varies with the square root of the distance would have seemed a violation of the fundamental canon of simplicity.

The idea of time as a precise measure of duration was evidently not so clearly grasped by the medieval scientist as the idea of space. Geometry and geometric instruments of precision were familiar, but medieval clocks were sorry affairs. The metricizing of time, urgently needed in order to develop the science of mechanics, was accomplished by Galileo, who, by 1610, had discovered that the law of falling bodies can be deduced by assuming that the velocity varies directly as the time. By thus introducing the concept of continuously varying instantaneous velocity, Galileo gave his contemporaries no end of trouble, since the concept implied the idea of motion with infinite slowness. It was easy enough to believe that a falling body passes through every single point of its continuous path to the ground, but not so easy to believe that the time of fall is also infinitely divisible. To seventeenth-century scientists this was as troublesome an idea as the concept of quantum-mechanical differential operators has been for many twentieth-century physicists. Galileo himself must have been troubled by the concept of continuously varying instantaneous velocity, for in his *Two New Sciences* he took pains to make it seem plausible by inventing ingenious but imaginary experiments.

By introducing mental experiments Galileo was invoking a method that has characterized physics ever since his time. He did perform actual experiments—with inclined planes, for instance—and much praise has been showered on him by undiscriminating admirers who have emphasized his zeal for experimentally testing all his conclusions. The melancholy fact, however, seems to be that Galileo's experiments were about as imprecise as most of the elementary physics lecture-table demonstrations of our own time, and Mersenne's complaint that he was unable to repeat Galileo's experiments and get the same results as Galileo is not surprising. But Galileo intended his demonstrations to be suggestive, not precise. He was not a great experimental physicist, but he was the founder of theoretical physics, and that is fame enough for any man.

Acoustics

Intensive historical study of the development of human ideas on acoustics was neglected until comparatively recent times. One reason for this neglect may be that the accepted root ideas on the origin, propagation, and reception of sound were proposed at a very early stage.[23] The ancient Greek philosophers were convinced that sound originates in the motion of the parts of bodies, that it is transmitted through the air by means of some undefined motions of the latter, and that this motion in the neighborhood of the ear produces the sensation of hearing. These ideas were vague enough, but they were much closer to what came to be accepted theory than were the ancient notions of the motion of large-scale objects or early theories of light and heat. In the latter two branches of physics especially, theory succeeded theory before the present point of view was attained. But in acoustics all that was needed was to elaborate and refine the basic ancient idea by appropriate mathematical analysis and to apply it to new phenomena as they were discovered. On its theoretical side in particular the history of acoustics thus tends to be merged in the general evolution of mathematical mechanics as a whole.

Many modern physicists believe that the essential physics of sound was worked out so long ago that it is no longer a physical subject but a branch of electrical engineering or possibly physiology. But these scientists' opinion has no more justification than the associated claim that because its fundamental notions were laid down early and have not changed significantly in the passage of time, acoustics has no history worthy of consideration. This opinion and claim can both be refuted by surveying in some detail the history of those parts of mechanics and other branches of physics that have a definite bearing on acoustical theory and acoustical practice.

The Production of Sound

It must have been observed from the very earliest times that when a solid body is struck in air, a sound is produced. The additional observation that under certain circumstances the sounds so produced are particularly agreeable to the ear furnished the basis for the creation of music, which must have originated long before the beginning of recorded history and which was of course also closely associated with the pleasant sounds that, again under favorable circumstances, may be emitted from the mouths of human beings, either directly to the ambient air or by means of a tube of appropriate shape. As far as we know from the available record, music was an art for millenia before its nature began to be examined in a scientific manner. It is usually assumed that the first Greek philosopher to study the origin of musical sounds was Pythagoras, who established his school in Crotone in southern Italy in the sixth century B.C. He is supposed to have been impressed by the facts that, of two stretched strings fastened at the ends, the note of higher pitch is emitted by the shorter string and that if one string has twice the length of the other, the shorter will emit a note an octave above the other. This account, which is probably legendary, is usually cited to provide a basis for the obsession that Pythagoras and his followers appeared to have for integral numbers as fundamental for the understanding of experience. It seems clear that the germ of the idea that pitch depends somehow on the frequency of vibration of the sound-producing object was in the minds of such Greek philosophers of the Pythagorean school as Archytas of Tarentum, in southern Italy, who flourished around 375 B.C. This point of view was also expressed in the writings on music of the Roman philosopher Boethius in the sixth century after Christ.[24]

For the modern scientific basis of this relation it has been customary to look to Galileo, who, at the very end of the "First Day" of his *Two New Sciences* (1638), included a remarkable discussion of the vibration of bodies.[25] Galileo began with the well-known observations on the isochronism of the simple pendulum and the dependence of the frequency of vibration on the length of the suspension. (In the former he made the mistake, perhaps excusable in his day, of concluding that the period of the pendulum is independent of the amplitude no matter how large the latter is.) He then went on to describe the phenomenon of sympathetic vibration, or resonance, in which the vibration of one body can produce a similar vibration in another distant body. He reviewed the common notions of the relation of the pitch of a vibrating string to its length and expressed the opinion that the physical meaning of the relation is to be found in the number of vibrations per unit time—what we now call the frequency. His view was confirmed by two observations. First, when the base of a glass goblet was fixed to the bottom of a large vessel and the goblet was filled with water almost up to its brim, he observed that the goblet could be made to vibrate and emit a sound if he rubbed its rim with his finger. At the same time he noticed that ripples ran across the surface of the water. And when, as occasionally happened, the note from the goblet rose an octave in pitch, the ripples in the water were divided in two—that is, the wave length was halved. The second observation was the result of an accident in which he happened to scrape a brass plate with an iron chisel in order to remove some spots from it. Once in a while the scraping would be accompanied by a sharp whistling sound of a definite musical character. When this happened, he always observed a long row of parallel fine streaks on the surface of the brass, equidistant from each other. He noticed further that the pitch of the whistling note could be increased by increasing the speed of scraping, in which case the separation of the streaks decreased.

Galileo was able to tune spinet strings with the aid of these chisel-scraping tones; he found that when the musical interval between two spinet strings was judged by ear to be a fifth, the average spacings between the lines on the brass plate for the corresponding scraping tones were in the ratio of 3 to 2. Galileo evidently understood the dependence of the frequency of a stretched string on its length, tension, and density; much of his knowledge was undoubtedly learned from predecessors. To explain why sounds of certain frequencies (those whose frequencies are in the ratio of two small integers) appear to the ear to combine pleasantly, whereas others not possessing this property sound discordant, he compared the vibrations of strings and pendulums. Galileo observed that a set of pendulums of different lengths, if set oscillating about a common axis and viewed in the original plane, present to the eye—at least to his eye—a pleasing pattern only if their frequencies are simply commensurable; otherwise they form a complicated jumble. This was a kinematic observation of great ingenuity and the basis of a suggestive analogy.

The history of science, like all history, is a matter of interpretation, and it is not surprising that Galileo's achievements in acoustics have been questioned. In his elaborate history of the mechanics of elasticity, Clifford

Truesdell has expressed the opinion that the importance of Galileo's contribution to the mechanics of vibration has been exaggerated. Truesdell has pointed out that though much of the material on general mechanics in Galileo's *Two New Sciences* (1638) dates from the early seventeenth century, when Galileo evidently first thought it out consistently, most of his results on vibrations were first published in *Two New Sciences*. Before its appearance several investigators had apparently come up with the fundamental ideas that Galileo expresses. The Frenchman Isaac Beeckman (1588–1637), who had evidently thought a good deal about the vibration of strings, published as early as 1618 some of his speculations manifesting his confidence in the relation between pitch and frequency, and giving arguments in its favor. (Beeckman is usually given credit for the initiation of Descartes into the study of physics.) Before Beeckman, Giovanni Battista Benedetti (1530–90) had published in Turin a work on musical intervals, in which he clearly stated his belief in the equality between the ratio of pitches and the ratio of frequencies of the vibrating motions corresponding to the production of the sounds. More elaborate were the studies of the French Franciscan friar, Marin Mersenne (1588–1648). In 1635 Mersenne published in his *Harmonicorum libri XII* some results he had obtained by experimentally observing the vibrations of a stretched string. He recognized that, other things being equal, the frequency of the vibration is inversely proportional to the length of the string, while it is directly proportional to the square root of the cross-sectional area. According to Truesdell, Mersenne anticipated Galileo in these important conclusions about vibrating strings.[26]

Later experimenters—like Robert Hooke (1635–1703), who is best known for his law of elasticity—tried to connect frequency of vibration with pitch by allowing a cogwheel to run against the edge of a piece of cardboard, a common lecture demonstration to this day.[27] But the man who made the most thoroughgoing pioneer studies of frequency in relation to pitch was undoubtedly a Frenchman, Joseph Sauveur (1653–1716), who also first suggested that the name acoustics be applied to the science of sound.[28] (*Acoustics* comes from the Greek word meaning hearing and is therefore to a certain extent appropriate; modern acoustics, however, far transcends the sounds we can actually hear.) Sauveur was aware of the significance of the beats that are observed when two organ pipes (or similar sound sources) of slightly different pitches are sounded together, and he actually used these beats to calculate the fundamental frequencies of two pipes that were adjusted by ear to be a semitone apart (so that their frequencies were in the ratio of 15 to 16). He found that when sounded together, the pipes gave six beats a second and, by treating this number as the difference between the frequencies of the pipes, reached the conclusion that their frequencies were respectively 90 cycles a second and 96 cycles a second. Sauveur also experimented with strings and in 1700 calculated by a somewhat dubious method the frequency of a given stretched string from the measured sag of the central point.

The English mathematician Brook Taylor (1685–1731), well known for his theorem on infinite series, was the first to provide a strictly dynamical solution of the vibrating string.[29] His solution, published in 1713, was based on an assumed curve for the shape of the string when vibrating in what we now call its fundamental mode (when all parts of the string are simultaneously on the same side of the equilibrium horizontal position). This curve was taken to be of such a character that every point would reach the horizontal position at the same time. From the equation of this curve and the Newtonian equation of motion, Taylor was able to derive a formula for the frequency of the fundamental vibration that agreed with the experimental law of Mersenne and Galileo. It is of particular interest to note that, as Truesdell has pointed out, this seems to be the first time that the Newtonian equation of motion ($F = ma$) was applied to an element of a continuous medium.[30]

Though Taylor treated only a special case and was clearly unable to progress to the treatment of the general string with all its modes because he lacked the calculus of partial derivatives, he did pave the way for the more elaborate mathematical techniques of the Swiss Daniel Bernoulli (1700–1782),[31] the Frenchman Jean d'Alembert (1717?–83),[32] and the Swiss Leonhard Euler (1707–83).[33] These men managed to set up the partial differential equation of motion of the vibrating string and to solve it in essentially the modern fashion. (The lack of adequate mathematical tools retarded the progress of the science of sound just as it held back the advance of mechanics in general. Unfortunately neither the fluxions of Newton nor the differentials of Leibniz were quite adequate for handling the motions of continuous media.)

To come back to the physical aspects of the problem of the vibrating string as a source of sound, it had already been observed by John Wallis (1616–1703)[34] in England and by Sauveur[35] in France that a stretched string can vibrate in parts, so that at certain intermediate points (Sauveur called them nodes) no motion takes place, whereas very violent motion takes place at other intermediate points he called loops. It was soon realized that such vibrations correspond to higher frequencies than the frequency associated with the simple vibration of the string as a whole without nodes and indeed that these frequencies are integral multiples of the frequency of the simple vibration. Sauveur called the associated emitted sounds the harmonic tones and the sound corresponding to the simple vibration the fundamental. The notation thus introduced around 1700 has survived to the present day. Sauveur noted also that a vibrating string could produce the sounds corresponding to several of its harmonics at the same time. The dynamical explanation of this was given by Daniel Bernoulli.[36] He showed that it is possible for a string to vibrate so as to cause a multitude of simple harmonic oscillations to be present at the same time and that each contributes independently to the resultant vibration, the displacement at any point of the string at any instant being the algebraic sum of the displacements associated with the various simple harmonic modes. He thus propounded the principle of the coexistence of small oscillations, also called the principle of superposition. Bernoulli tried to give a proof of the principle but did not succeed; his grasp of mathematics was not so great as his understanding of physical ideas. The real significance of the superposition principle, that the partial differential equation governing the motion of the ideal frictionless string is linear, was pointed out almost immediately by Euler.[37] With this understanding, the superposition principle can be proved as a theorem.

The history of the theory of the vibrating string in the eighteenth century consists of a series of controversies in which ingenious investigators like Bernoulli, Euler, and D'Alembert argued vehemently with each other. They took their work very seriously and did not hesitate to cast harsh aspersions on each other. The mathematics needed for the description of the motion of continuous media, so fundamental for the progress of acoustics as an exact science, was being born during this period, and the travail was not easy. The standard texts tend to gloss over the unpleasant things that even the great scientists of the time said about each other in their correspondence and articles, and the serious errors that they often made. The possibility of expressing any arbitrary function—such as the initial shape of a vibrating string in terms of an infinite series of sines and cosines, implied by the superposition theorem—was hard to accept in terms of mid-eighteenth-century mathematics. It was not until 1822 that, in his analytical theory of heat, Jean Baptiste Joseph Fourier (1768–1830) based his celebrated theorem on this type of expansion, with consequences of the greatest value for the advancement of acoustics.[38]

Among the eighteenth-century mathematicians who tackled the problem of the vibrating string was Joseph Louis Lagrange (1736–1813), an Italian from Turin who spent most of his active career in France. He was the author of *Mécanique analytique* (1788), a treatise in which mechanics was reduced to a branch of mathematical analysis; in the Preface he boasted that he had included no figures, for they were unnecessary. (The reader who has studied theoretical physics will recall generalized co-ordinates and Lagrange's equations.) In an extensive memoir presented to the Turin Academy in 1759, Lagrange adopted what he claimed was a different and novel approach to the string problem.[39] He assumed a string to be composed of a finite number of equally spaced particles identical in mass and tied together by equal segments of stretched, weightless string. He then solved the problem of the motion of this system as a dynamical system with many degrees of freedom and established the existence of a number of independent frequencies equal to the number of the particles. When he passed to the limit and allowed the number of particles to become infinitely great and the mass of each correspondingly small (so that the product equalled the finite mass of the string), these frequencies were found to be precisely the harmonic frequencies of the stretched continuous string. Lagrange claimed that his device avoided the analytical difficulties associated with the motion of the continuous string and that he had made a decisive advance. Euler, however, had already, in 1744, solved the mechanical problem of the motion of n particles on a string, where n is any integer, though he had not been successful in the passage to the limit.[40] Moreover, as Truesdell has pointed out, Lagrange's passage to the limit was mathematically faulty and, to be made rigorous, demanded essentially the same kind of mathematical assumptions that Lagrange objected to in the analysis of his contemporaries Bernoulli, Euler, and D'Alembert.[41] Nevertheless, Lagrange's method was adopted by Lord Rayleigh in *The Theory of Sound* and has found its way into most modern texts in mechanics and acoustics. It is not the most direct way to handle the vibrating string, and undoubtedly Lagrange exaggerated the significance of his accomplishment. But his method has the merit of variety, and this is important in science: the more ways in which the same phenomenon can be looked at, the better it can be grasped.

D'Alembert is usually credited by historians with having been the first, in 1747, to develop the partial differential equation of the vibrating string in the form now referred to as the wave equation. He also found the solution of the equation in the form of waves traveling in both directions along the string. From this point of view the vibrations of the string are due to a combination of such traveling waves forming standing or stationary waves. There was much controversy about the meaning and validity of D'Alembert's theory, just as there is controversy today over the development of theories of relativity and quantum mechanics.

The vibrating string, important as it was, did not absorb all the attention of eighteenth-century scientists. They were interested in other sound-producing motions as well. In Lagrange's 1759 memoir, for example, there is a treatment of the sounds produced by organ pipes and other musical wind instruments. The basic experimental facts were already known, and —although the boundary conditions gave some trouble, as they still do— Lagrange was able to predict theoretically the approximate harmonic frequencies of closed and open pipes. Euler also made great contributions, the magnitude of which has only recently been appreciated, to the field of the harmonic frequencies of closed and open pipes. In 1727, long before Lagrange's memoir, Euler, who was only twenty years old and at that time particularly interested in musical instruments like the flute, set forth in essentially modern form the basic features of the overtones of pipes.[42] Around 1759 both Euler and Lagrange were deeply engrossed in the subject of sound oscillations in tubes and corresponded copiously. In 1765 Euler produced an elaborate treatise on fluid mechanics, one section of which was entirely devoted to sound waves in tubes.[43] It is difficult today to appreciate the tremendous zeal with which such problems were tackled and solved by the great mechanically minded mathematicians of this era, which may justly be called the golden age of mathematical physics.

The mathematical scientists of the eighteenth century realized that other solid bodies beside strings emit sound when disturbed. For example, they were of course familiar with bells, and a vast amount of empirical knowledge must have accumulated about such sound sources. But the successful application of mathematical methods to the vibrations of metal bars, plates, and shells demanded a knowledge of the relation between the deformation of the solid body and the impressed deforming force. Fortunately, this problem had already been tackled and solved in its simplest form by Robert Hooke, who in 1660 discovered, and in 1675 announced, his law relating the stress and strain for bodies undergoing elastic deformations.[44] Hooke's law states that within what is called elastic limit the strain of an elastic body (the fractional increase in length for a linear rod or bar) is directly proportional to the stress (the force per unit area of cross section of the rod or bar in the direction of the stretch). This law forms the basis for the whole mathematical theory of elasticity, including elastic vibrations that give rise to sound. Its application to the vibrations of bars that are supported and clamped in various ways appears to have been made as early as 1734/35 by Euler and by Daniel Bernoulli.[45] The mathematical methods they used, which were later systematized and extended by Lord Rayleigh in *The Theory of Sound,* fundamentally involved beginning with the expression for the energy of a deformed bar and using the variational technique that leads to the well-known equation of the fourth order in space derivatives.

The corresponding analytical solution for the vibrations of a solid elastic plate proved much more difficult and came much later, though much useful experimental information was obtained in the latter part of the eighteenth century, when the German scientist Ernst Florens Friedrich Chladni (1756–1827) described his method of using sand sprinkled on vibrating plates to show the nodal lines (lines of zero displacement).[46] These Chladni figures, which have long been recognized as things of great beauty, could in a general way be accounted for by considerations similar to those explaining the existence of nodes in a vibrating string. The exact forms, however, defied analysis for many years, even after the publication of Chladni's classic *Die Akustik* in 1802. In 1815 a prize of 3,000 francs provided by the emperor Napoleon for an award by the Institute of France for a satisfactory mathematical theory of the vibrations of plates was won by the mathematician Sophie Germain (1776–1831). But while she produced the correct fourth-order equation, her choice of boundary conditions proved to be incorrect, and it was not until 1850 that Gustav Robert Kirchhoff (1824–87) furnished a more accurate theory.[47] Modern technology, with its concern for the vibrations of such things as airplane fuselages, still supports active research on the vibrations of plates and solid shells of various shapes.

The first solution of the analogous problem of the vibrations of a flexible membrane is usually attributed to the French mathematician Siméon Denis Poisson (1781–1840).[48] He, however, failed to complete the case of the circular membrane; this was not accomplished until 1862 by Alfred Clebsch (1833–72).[49] As Truesdell has pointed out, the attribution to Poisson neglects a very important work seventy years earlier in which Euler derived the appropriate partial differential equation for the vibrating membrane and expressed it properly for both rectangular and circular shapes.[50] But Euler made a curious slip in assigning boundary conditions and failed—by a factor of 2—to get the correct answer for the normal modes. Poisson, apparently not familiar with Euler's work, was able to get the correct answer for the rectangular membrane.

It is significant that much of the theoretical work on vibration problems during the nineteenth century was carried out by persons who called themselves mathematicians. This was natural, for much of the application of the mathematics of that period was to physical problems of this kind; but it was also unfortunate because the choice of boundary conditions did not always reflect experimentally realizable situations. The ability to excite vibrations with arbitrary frequency over a wide range in media of arbitrary nature, size, and shape had to await the development of electroacoustics, largely a product of twentieth-century research.[51] It is true that electrical oscillations and oscillating electric circuits were discovered and invented in the middle of the nineteenth century, but practical methods for coupling them to mechanical systems to make them produce mechanical oscillations were not developed until after 1900. As long as tuning forks remained the only practical frequency standards for sound sources, no great progress in applying the mathematical theory of acoustics to practical cases could be expected; however, the basic physical principles for doing so were well known in the nineteenth century. The force on a current-carrying conductor in a magnetic field had been discovered in the 1820's, and the piezoelectric effect emerged from the experiments of the Curie brothers in 1880.[52] (The piezoelectric effect is the property, displayed by some crystals —notably quartz—of having electric charges appear on their faces when subjected to mechanical stresses of various kinds and, conversely, the property of changing dimensions—exhibiting strain—when placed in an electric field.) It was early recognized that the piezoelectric effect could be used to develop both controlled sources and receivers of sound waves, but the actual exploitation of the effect for this purpose did not take place until the second decade of the twentieth century. The same situation prevailed with the magnetostriction effect (the tendency of magnetizable materials to change dimensions when placed in a magnetic field), which was discovered by James Prescott Joule (1818–89) in 1842.[53] Only the advent of the vacuum-tube oscillator and amplifier made it possible to use these effects to produce and receive sound at all frequencies and intensities on a precision basis.

Among the most important producers of sound are, of course, the vocal cords, and it is a curious fact that though examples of this kind of sound production are in many ways the most obvious of all, little attention was paid to them during the historical evolution of acoustics just surveyed. The very obviousness of speech may well have caused those who were concerned with the physically objective aspects of sound to neglect it. Speech, after all, seemed closer to language and therefore in the province of the philologists and etymologists, and the basic mechanism of human speech—the combination of vibrating vocal cords and mouth cavities—was considered to be material for anatomists and physiologists. Nevertheless, as early as 1629 the Englishman W. Babington observed the motions of the vocal cords by means of light reflected from mirrors in the mouth. This was the beginning of the development of what came to be called the laryngoscope, finally perfected by the Czech physiologist Johann Czermak in 1857. Some eighty years later movies of the vocal cords were made at the Bell Telephone Laboratories.

The nature of the vowel sounds of speech and singing was first thoroughly investigated in 1860 by Hermann von Helmholtz (1821–94), who used the resonators that bear his name.[54] Investigations by Sir Charles Wheatstone (1802–75) in 1837 had led to the development of a harmonic theory for the production of vowel sounds. According to this theory, the vocal cords vibrate so as to produce both a fundamental frequency and numerous harmonics; it was assumed that the vibrations when communicated to the air are reinforced by resonance in the mouth cavities. Another theory, proposed by W. T. Willis in England in 1829, assumed that the vowel sound is not caused by continuous vibration of the vocal cords but rather by puffs of air emitted by them: these transient puffs set the air in the mouth cavities in vibration, and the resonance there gives the emitted sound its characteristic quality. Helmholtz later pointed out that both ideas have elements of correctness, and modern research has confirmed this opinion. Some early workers held that the whole oral cavity acts as a single resonator, while others (notably Alexander Graham Bell and Helmholtz himself) believed that vowel sounds depend on two characteristic resonances, corresponding to the action of the mouth as a double resonator (two resonators coupled together). As Helmholtz pointed out in his 1862 treatise, the knowledge that the oral cavities can be tuned to different frequencies goes back to the early seventeenth century, long before instrumental study of the phenomenon was highly developed.

The Propagation of Sound

From the earliest recorded observations, there has been general agreement that sound is conveyed from one point in space to another through some activity of the air. Aristotle emphasized that actual motion of air is involved, and it is possible to read into the descriptions in his treatise *De anima* and in *De audibilibus*, the treatise often ascribed to him, the notion that sound is due to compressional waves in air.[55] However, the problem of interpretation that plagues the history of science makes it difficult to be sure that Aristotle and his contemporaries had really grasped the idea that in the propagation of sound the air does not move as a whole (as a stream) in the direction of propagation. It seems clear that the Roman architect and engineer Vitruvius in the first century B.C. had an adequate grasp of the wave theory of sound from the analogies he drew with surface waves on water.[56] However, since in the transmission of sound the air certainly does not appear to move, it is not surprising that other, much later, philosophers denied these views of Aristotle and Vitruvius.

Even during the lifetime of Galileo, Gassendi, in his revival of the atomic theory, attributed the propagation of sound to the emission of a stream of very small, invisible particles from the sounding body that after moving through the air, are able somehow to affect the ear.[57] Otto von Guericke (1602–86), in his 1672 treatise *Nova experimenta de vacuo spatio*, expressed great doubt that sound is conveyed by a motion of the air, observing that it is transmitted better when the air is still than when there is a wind. Moreover, having tried, around 1650, the experiment of ringing a bell in a jar that was evacuated by means of an air pump, he had still been able to hear the sound. The first to try the bell-*in-vacuo* experiment was apparently the Jesuit Athanasius Kircher (1601–80), who described it in his book *Musurgia universalis* (1650) and concluded from his observation that air is not necessary for the transmission of sound. Undoubtedly, both Guericke and Kircher had failed to avoid transmission through the walls of the vessel or to obtain an adequate vacuum.

In 1660 Robert Boyle (1627–91) repeated the experiment in England with a much-improved air pump and more careful arrangements, and finally observed the now well-known decrease in the intensity of the sound as the air is pumped out.[58] He concluded that the air is definitely a medium for the transmission of sound, though presumably not the only one. Actually, the observed decrease in the intensity of the sound is due not so much to the failure of the low-pressure air to transmit sound as to the increasing difficulty of getting the sound out of the bell (or other sound source) into the air and then out again from the air to the glass container. The impedance mismatch between source and surrounding fluid medium becomes greater as the density of the medium decreases. The experiment does, however, demonstrate a very important connection between the source and the medium in acoustic propagation, and the modern theory of sound implies that some material medium is necessary for acoustic transmission.

Granted that air is a sound-transmitting medium, the question at once arises of how rapidly the propagation takes place. As early as 1635 Gassendi, while in Paris, made measurements of the velocity of sound in air, using firearms and assuming that the light of the flash is transmitted instantaneously.[59] His value came out to be 1,473 Paris feet (478.4 meters) per second. (Gassendi, of course, did not use the metric system, which was not introduced until the time of the French Revolution.) Somewhat later, by more careful measurement, Mersenne showed that Gassendi's figure was too high; Mersenne's value was 1,380 Paris feet (about 450 meters) per second. Gassendi noted the important fact that the velocity is independent of the pitch of the sound and thus discredited the view of Aristotle, who had taught that high notes are transmitted faster than low notes. On the other hand, Gassendi made the mistake of believing that wind has no effect on the measured velocity of sound. In 1650 two Italians, Giovanni Alfonso Borelli (1608–79) and his colleague in the Accademia del Cimento of Florence, Vincenzo Viviani (1622–1703), tried the same type of experiment and obtained 1,077 Paris feet (350 meters) per second.[60]

All of these measurements suffered from lack of reference to the temperature, humidity, and wind velocity. Though the Englishman William Derham (1657–1735) made extensive measurements of the velocity of sound in the early part of the eighteenth century in which he concluded that the velocity is independent of all environmental conditions except wind,[61] his results were shown to be mistaken by the Italian Giovanni Ludovico Bianconi (1717–81), who in 1740 demonstrated that the velocity of sound in air increases with the temperature.[62] The first open-air measurement that can be considered at all precise in the modern sense was probably that carried out under the direction of the Academy of Sciences of Paris in 1738, using a cannon as the source of sound. When reduced to 0°C, the result in modern units was 332 meters per second. Careful repetitions during the two succeeding centuries gave results differing from this value by less than 1 per cent. The best modern value (1942) is 331.45 ± 0.05 meters per second under standard conditions of temperature and pressure.[63] This is a tribute to the care with which these Paris academicians carried out their work—very few early physical measurements have stood the test of time as well as these of the velocity of sound in air.

The problem of the measurement of the velocity of sound in solid media was tackled by Chladni, whose investigations of the nodal lines in vibrating plates have already been mentioned. He used similar means to study propagation in metal rods, and by measuring internodal distances he was able to calculate sound velocity in such specimens. In 1808 the French physicist Jean-Baptiste Biot (1774–1862), best known for his work in optics, made actual measurements of the velocity of sound in an iron water pipe by direct timing.[64] The pipe was nearly 1,000 meters long, and by comparing the times of arrival of a given sound through the metal and through the air, respectively, he established that the velocity of the compressional wave in solid metal is many times greater than that in air. This was indeed to be expected from the very much greater elasticity of metal in comparison with air. Biot's experimental values agreed in order of magnitude with those of Chladni.

The first serious attempt to measure the velocity of sound in a liquid like water was apparently that of the Swiss physicist Daniel Colladon (1802–93), who was assisted by the mathematician Charles Sturm (1803–55). The Academy of Sciences in Paris had announced as the subject of its prize competition for 1826 the measurement of the compressibility of the principal liquids. Colladon entered the competition and successfully measured the static compressibility of water and some other liquids. He must have been fascinated by its relatively low value and the correspondingly large magnitude of its reciprocal, the bulk modulus. He was, of course, aware of the theoretical relation between the compressibility and the speed of sound. Tests to check the accuracy of his compressibility measurements by acoustic velocity were carried out in Lake Geneva in November, 1826, and the published results won the prize.[65] The compressibility of water as computed from the velocity of sound turned out to be very close to the statically measured value. The story of Colladon and Sturm's study is found in Colladon's autobiography, in which the man who did the work tells how he did it and includes such homely details as the troubles he encountered in carrying the powder needed for his light flashes across the frontier between Switzerland and France.[66] The average velocity found in this measurement was 1,435 meters per second at 8°C.

Though the propagation of sound through air had already been compared with the motion of ripples on the surface of water, the first attempt to theorize seriously in mathematical form about a wave theory of sound was apparently made by Newton, who, in the second book of his *Principia Mathematica* (Propositions 47, 48, and 49), compared the transmission of sound with pulses produced when a vibrating body moves the adjacent portions of the surrounding medium and these in turn move those adjacent to themselves, and so on. Newton then went on to make some rather arbitrary assumptions, the principal one being that when a pulse is propagated through a fluid, the particles of the fluid always move in simple harmonic motion—"are always accelerated or retarded according to the law of the oscillating pendulum" (Proposition 47). He proved the theorem that if this is true for one particle, it must be true for all adjacent ones. The end result is that the velocity of sound is equal to the square root of the ratio of the atmospheric pressure to the density of the air.

As was to be expected, Newton's "derivation" was subjected to much criticism by the natural philosophers of Continental Europe; among the critics were Euler, Johann Bernoulli (younger brother of Daniel), and Lagrange. In 1727, in his remarkable work *Dissertatio physica de sono,* Euler set forth with the greatest clarity his ideas of the physical principles underlying sound propagation as well as sound production and attacked Newton's method as being entirely too specialized. He presented an expression for the velocity of sound in air very close to Newton's, though it must be admitted that Euler's method is not clear. In a later treatise (*De propagatione pulsuum per medium elasticum,* 1748), Euler developed Newton's theory in much clearer analytic form and obtained Newton's result.

The problem of sound propagation could not be completely resolved until the wave equation for sound waves in a fluid could be set up and solved. D'Alembert had first derived this equation for a continuous string in 1747; at that time he commented on the fact that it should be possible to apply the same equation to sound waves. However, he did not get far with the details, which were worked out later by Euler. Euler laid the foundation for the theory of the propagation of sound waves in air in three great memoirs to the Berlin Academy in 1759, but it was Lagrange who, in memoirs to the Turin Academy at about the same time, revised Newton's reasoning and generalized it to the case of sound waves of arbitrary character as distinct from simple harmonic waves. Euler arrived at the same result as Newton for the velocity of sound in air—evidence either of Newton's genius or of his good luck.

It is well known that when the relevant data for air at 60°F (15.5°C) are substituted into the Newtonian equation for the velocity of sound ($c = \sqrt{p/\rho}$, where p is the gas pressure and ρ the corresponding density), the result is 945 feet (about 310 meters) per second. Though this is definitely lower than the experimental results of Paris that have been mentioned, Newton at first thought that the order-of-magnitude agreement was satisfactory. However, when more accurate measurements confirmed the higher value, Newton revised his theory in the second edition of his *Principia* (1713) to try to produce better agreement with experiment. What his line of thought was is not at all clear, but he evidently felt that some correction must be made for the impurity of the actual air.

Nothing more seems to have been done on the problem of the velocity of sound until 1816, when Pierre Simon de Laplace (1749–1827) suggested that an error had been made in the Newtonian and Lagrangian determinations in using the pressure itself for the volume elasticity of the air (the reciprocal of the compressibility); this is equivalent to assuming that the elastic motions of the air particles take place at constant temperature.[67] In view of the rapidity of the motions involved in the passage of the sound wave, it seemed more reasonable to Laplace to suppose that the compressions and rarefactions follow the adiabatic law on the assumption that heat does not have a chance to flow out of the compressed region before compression gives way to rarefaction. But in this case the adiabatic elasticity is higher than the isothermal value in the ratio γ, where γ is the ratio of the specific heat of the gas at constant pressure to the gas's specific heat at constant volume. According to this line of reasoning, Newton's expression should be changed to $c = \sqrt{\gamma p/\rho}$; since γ is always greater than 1, the newly calculated velocity of sound would necessarily be greater than the old and, therefore, closer to the experimental value. When Laplace put forth his theory in 1816, the existence of two specific heats of a gas was recognized, but the value of γ was not known very precisely. Using the value of 1.5 for air as obtained in early measurements of specific heat, Laplace found c equal to 345.9 meters per second at 6°C, compared with the best experimental value then available of 334.18 meters per second for this temperature. This was close enough for Laplace to feel he was on the right track. He returned to the problem later and included a chapter on the velocity of sound in his *Mécanique céleste* (1825). In 1819 Nicolas Clément and Charles-Bernard Désormes performed the classical experiment on the determination of γ and found the value of 1.35, leading to 332.9 meters per second for the sound velocity at 6°C. Some years later the more accurate value of 1.40 led to complete agreement between Laplace's theory and experiment. This theory is now so well established that it is common practice to determine γ for various gases by precision measurements of the velocity of sound.

The latter half of the eighteenth century and the first quarter of the nineteenth witnessed numerous attempts to theorize about waves in continuous media. These attempts were based largely on D'Alembert's solution for the wave equation, the equation that states, in effect, that the second time derivative of the quantity that "waves" is equal to the second space derivative of this same quantity multiplied by the square of the wave velocity. Much attention, for example, was paid to waves on the surface of liquids like water. This work had value in connection with acoustics only to the extent that it led to increased confidence in the applicability of the wave equation to sound propagation in fluids. By 1800 the solution of the equation for aerial sound propagation in tubes subject to the boundary conditions at the ends had been pretty well established, and the predicted harmonic frequencies (normal modes or partials) checked with experiment with reasonable accuracy. There were, however, puzzling discrepancies in detail that were not to be fully cleared up until the "end corrections" were understood some half-century later. Experimental techniques for sound measurement in tubes remained rather crude for a long time; it was not until 1866 that August Kundt (1839–94) developed his simple but effective method of dust figures for studying experimentally the propagation of sound in tubes and in particular for measuring sound velocity in air and other gases from standing wave patterns (nodes and loops).[68]

In the meantime, in a memoir published in 1820, Siméon Denis Poisson attacked the much more difficult problem of the propagation of a compressional wave in a three-dimensional fluid medium.[69] Three years earlier Poisson had devised the most elaborate theory thus far of the transmission of sound in tubes, which included the theory of stationary air waves in tubes of finite length, both open and closed.[70] He even considered the possibility of an end correction to take care of the fact that the condensation (the fractional change in density due to the sound wave) cannot be considered as precisely zero at the open end, with the result that the observed resonance frequencies correspond to a length slightly greater than the actual geometrical length of the tube. (Helmholtz treated this whole problem more thoroughly in 1860.) The special case of an abrupt change in cross-section was also studied by Poisson, along with the reflection and transmission of sound at normal incidence on the boundary of two different fluids. Much modern work of practical significance was anticipated in this great 1820 memoir of Poisson.

The more difficult problem of the reflection and transmission of a plane sound wave incident obliquely on the boundary of two different fluids was solved by the self-taught Nottingham genius George Green (1793–1841) in 1838. His memoir emphasized the refraction of sound and stressed both the similarities and differences between the reflection and refraction of sound and light: that sound waves in ideal fluids, being strictly compressional, are longitudinal, whereas light waves are transverse; that light waves can therefore be polarized, while sound waves in fluids cannot; and that, on the other hand, elastic waves in an extended solid can be both longitudinal and transverse, or more accurately, irrotational and solenoidal. Poisson's 1829 study of isotropic elastic media also shows an understanding of reflection and refraction, but at that time they did not seem to have much significance for acoustics.[71] They did, however, have a very important bearing on the elastic-solid theory of light, which was actively pursued during the middle decades of the nineteenth century. Today this early work of Green and Poisson has new and great significance because of interest in the propagation of elastic waves in structures like airplane fuselages and space missiles. Its importance for modern geophysics (seismological waves) is obvious.

So far in this historical résumé of the propagation of sound it has been tacitly assumed that the disturbance in the material medium being propagated as a sound wave (the excess density or pressure in a fluid) is very small compared with the equilibrium value. In this case the equation for wave propagation is linear. This is the type of equation to which the eighteenth-century investigators in acoustics gave their full attention. That its solution gives only an approximation to the actual sound transmission for relatively large disturbances was finally realized in the nineteenth century. However, Euler had already come close to what is called the finite-amplitude wave equation in a 1759 memoir on the propagation of sound in which he worked out the equation of motion of a thin slice of air subject to pressure forces on its two sides.[72] His physics was in this case impeccable, and he would have obtained the precise result of nineteenth-century research had he not made an unaccountable algebraic error. At any rate he realized that the normal linear wave equation, containing only second derivatives of the wave-displacement function with respect to space and time, must be corrected by the inclusion of a nonlinear term whenever the gradient of the displacement is an appreciable fraction of unity.

Nonlinear acoustic wave propagation was not taken up again seriously until around the middle of the nineteenth century, when the German mathematician Georg F. B. Riemann (1826–66) and the British mathematician and physicist Samuel Earnshaw (1805–88) more or less independently investigated certain special cases.[73] Their results showed that in nonlinear propagation the velocity of propagation depends on the amplitude, so that it is only under very special conditions that a nonlinear wave of permanent type can be realized. Some understanding of this situation had previously been reached by Poisson.[74] All this work led up to the theory of shock waves developed by Sir George Gabriel Stokes, James Challis, William John Macquorn Rankine, H. Hugoniot, and Lord Rayleigh, among others.[75] Nonlinear acoustics has assumed great importance in its twentieth-century development.

The Reception of Sound

The human ear has had a greater influence on the development of acoustics than human speech; for a long time the reception of sound was studied largely in connection with the behavior of the ear as a sound receiver. The ear is remarkably versatile and sensitive. It has been established that its normal threshold of hearing is a sound intensity of about 10^{-16} watt per square centimeter, or 10^{-9} erg per square centimeter per second. If the area of the normal ear drum is about 0.66 square centimeter, an average mechanical energy flow of only 6.6 times 10^{-10} erg per second can produce the sensation of sound. A harmonic sound in the audible range of frequency will be identified if its duration is of the order of 0.1 second. Thus acoustic energy of the order of 6.6 times 10^{-11} erg is sufficient to excite identifiable sound in the ear. In terms of energy the ear turns out to be fully as sensitive as the eye. Many elaborate investigations of the anatomy of the ear have been made over the past century, and its acoustical behavior has been studied intensively; nevertheless, no completely acceptable theory of audition has emerged. Precisely how we hear still remains a puzzling problem in modern psychophysics.

After the relation between pitch and frequency had been established, the next task was to determine the frequency limits of audibility. The French physicist Félix Savart (1791–1841), using fans and rotating toothed wheels in investigations around 1830, placed the minimum-audible frequency at 8 vibrations per second and the upper limit at 24,000 vibrations per second[76] (now usually referred to as cycles per second, or "hertz" after the German physicist Heinrich Hertz [1857–94], whose studies on electromagnetic waves were epoch-making). Other investigators—among them Ludwig F. W. A. Seebeck (1805–49) who is not to be confused with Thomas Johann, the discoverer of the thermoelectric effect; Karl Rudolf Koenig (1832–1901); Jean-Baptiste Biot; and Helmholtz—obtained values for the lower limit ranging from 16 to 32 cycles per second.[77] Their results emphasize the role that individual differences play in hearing, differences that are even more noticeable in the case of the upper-frequency limit of audibility. The latter not only varies considerably from individual to individual, but usually decreases with the age of an individual. It has become clearer in comparatively recent times that the values in each case depend on the intensity. The most elaborate studies on audibility made during the nineteenth century were those of Koenig, who, before the era of electroacoustical sources, devoted a lifetime to the design and production of precision sources of sound of controlled frequency, such as tuning forks, rods, strings, and pipes.[78] Koenig was also responsible for the electrically driven tuning fork.

Around 1870 the closely related problem of minimum-sound intensity necessary for audibility (the auditory threshold) was apparently first studied by August Toepler (1836–1912) and the great Viennese theoretical physicist Ludwig Boltzmann (1844–1906), who is best known as one of the creators of the statistical theory of gases. By an ingenious use of optical interference these collaborators were able to measure the maximum change in density (or effectively, the maximum condensation) in a just-audible sound wave. Their experimental results led to an audible threshold of about 10^{-11} watt per square centimeter—considerably in excess of that obtained by modern methods but at any rate suggestive of the great sensitivity of the human ear.[79]

In 1843 Georg Simon Ohm (1787–1854), the author of the law of electric currents, put forward a theory of audition, according to which all musical tones arise from simple harmonic vibrations of definite frequency and the particular quality or timbre of actual musical sounds is due to combinations of simple tones of commensurable frequencies.[80] He held, moreover, that the human ear is able to analyze any complex note into a set of simple harmonic tones, in terms of which it may be expanded mathematically by means of Fourier's theorem. This theorem states that (subject to certain specific mathematical restrictions) any arbitrary function of a variable t can be expanded in a convergent series of circular functions whose arguments are integral multiples of a fundamental frequency. If the arbitrary function is itself periodic in time, it can be represented in this way for all values of the variable t, whereas if the function is not periodic, it can be so represented only over a finite time interval. This theorem has been of great value in the analysis of sounds of all sorts.

Ohm's law stimulated much research in what has come to be called physiological and psychological acoustics—the acoustics of hearing. The most important nineteenth-century research in this field was that of Hermann von Helmholtz, who formulated the first elaborate theory of the mechanism of the ear—the so-called resonance theory.[81] According to Helmholtz' theory, by which he was able to justify theoretically Ohm's law, the various elements of the basilar membrane in the cochlea resonate to certain frequencies in the sound falling on the ear. Helmholtz became interested in the mechanical phenomenon of resonance and in the course of his investigations invented the special type of sound resonator since known by his name. The Helmholtz resonator is simply a spherical chamber with an orifice. When a harmonic source of sound of appropriate frequency is brought close to the opening, if the sizes of the chamber and the orifice are just right, the sound will be much amplified by the vigorous oscillatory motion of the air in the orifice. A large chamber resonates to a low-frequency or low-pitch tone and a small chamber to high frequency or high pitch. Such resonators have had wide use in modern acoustical research and applications. Helmholtz showed that when two tones of different frequencies are directed at an asymmetrical vibrator, the resulting vibration will contain frequencies that are the sum and difference of the original ones, and many other linear combinations of the original frequencies will occur. He speculated that the eardrum is such an asymmetrical vibrator and predicted human ability to detect such summation and difference tones. This prediction has been verified. Helmholtz' pioneer researches laid the groundwork for all subsequent research in the field of audition. An experimental and theoretical genius, he was one of the greatest physicists of the nineteenth century.

The reception of sound by the ear in enclosed spaces has great practical application, and for this reason considerable attention has been given by physicists to the development of what has come to be called room or architectural acoustics. It was recognized early that some rooms are not satisfactory for good hearing, and various devices were used to overcome the difficulties. These were at first simple geometrical contrivances like sounding boards and other reflectors. But in 1853 a Boston physician, J. B. Upham, wrote several papers on the more important problems involved in the reverberation or multiple reflection of sound from all surfaces of a room. He also showed how the reverberation time could be reduced by installing fabric curtains and upholstered furnishings. In 1856 the distinguished American physicist Joseph Henry (1797–1878), who became the first secretary of the Smithsonian Institution, made a study of auditorium acoustics that reflects a clear understanding of all the factors involved, though all of his suggestions were qualitative.[82] In spite of these early moves in what is now recognized to have been the right direction, acoustics was completely neglected by architects, and during the latter half of the nineteenth century attempts were often made to correct gross acoustical defects in rooms by absurdly inadequate devices. (One method was to string up wires.) The modern quantitative foundation of architectural acoustics dates from the work of the physicist Wallace C. Sabine (1868–1919), of Harvard University, who, in 1900, hit upon the law connecting the reverberation time in a room (the time taken for any initially built-up sound intensity to decay to any arbitrarily chosen fraction of its original value, say the one-millionth part) with the volume of the room and the amount of acoustic absorbing material in it.[83] This made applied architectural acoustics possible in the sense that any room could now be designed so that speech could be heard satisfactorily. To a certain extent this became true for music also, though subtle psychological factors enter that can be troublesome.

Special devices for amplifying sound received by the ear have a long history. Horns, for example, are of great antiquity, but it is uncertain when the suggestion was first made that they might be used to improve the reception of sound. In 1650, however, Athanasius Kircher designed a parabolic horn for use as a hearing aid as well as a speaking trumpet; he evidently was aware of the importance of the flare in the amplification of both received and emitted sound. Robert Hooke experimented with ear trumpets and is even supposed to have suggested the possibility of a device to magnify the sounds of the body.[84] In 1817–19 the French physician René Laënnec (1781–1826) invented the stethoscope and initiated its use for clinical purposes.[85] A similar instrument was developed in 1827 by the English physicist Sir Charles Wheatstone. He called it a microphone, a name now applied to an electromechanical device (one in which motion produced by sound is made to induce electric currents) for the reception of sound.[86] Electroacoustics, without which modern experimental acoustics could not exist, had then hardly been thought of.

All through the historical development of physics there is a recurring tendency to reduce the observation of physical phenomena, and especially experimental measurements, to something that can be seen. Practically all physical measurements involve this principle and employ a pointer or a spot of light moving on a scale. It was inevitable that attempts would be made to study sound phenomena visually; this was, of course, necessary for the investigation of sounds whose frequencies lie above the range of audibility of the ear (ultrasonic radiation). One of the first moves in this direction was the observation by the American physicist John LeConte (1818–91) that musical sounds can produce fluctuations in a gas flame if the pressure of the gas is properly adjusted.[87] The "sensitive flame," as it later came to be called, was developed to a high degree of precision by the English physicist John Tyndall (1820–93), who used it to detect high-frequency (inaudible) sounds and to study the reflection, refraction, and diffraction of sound waves.[88] It still provides an effective lecture demonstration, but for practical applications it has been superseded in the twentieth century by various types of electrical microphones coupled to the cathode-ray oscilloscope.

In the endeavor to make the form of a sound wave visible, Koenig, around 1860, invented the manometric flame device, which consists of a box, one side of which is a flexible membrane. Gas flows through the box to a burner, and when sound waves strike the membrane, the alternating changes in pressure produce corresponding fluctuations in the flame that can be made visible by reflecting the light of the flame from a rapidly rotating mirror.[89] Another attempt to visualize sound waves was made by the nineteenth-century French proofreader, editor, and amateur scientist Edouard-Léon Scott de Martinville in 1857 by means of his "phonautograph," in which a flexible diaphragm at the throat of a receiving horn was attached to a stylus, which in turn touched a smoked rotating-drum surface and traced out a curve corresponding to the incident sound.[90] This was the precursor of the phonograph. An equally ambitious attempt along similar lines was made by Eli Whitney Blake (1836–95), the first Hazard Professor of Physics at Brown University, who, in 1878, made a microphone by attaching a small metallic mirror to a vibrating disc at the back of a telephone mouthpiece.[91] By reflecting a beam of light from the mirror, Blake succeeded in photographing the sounds of human speech. Such studies were advanced further by the American physicist Dayton Clarence Miller (1866–1941), who invented a similar instrument, the "phonodeik," and made very elaborate photographs of sound-wave forms.[92] These devices of Scott, Blake, and Miller were of course the predecessors of the cathode-ray oscilloscope that is so useful in modern acoustical research.

Lord Rayleigh and Modern Acoustics

The publication of Lord Rayleigh's *The Theory of Sound* in 1877 marked both the end of what may be called the classical era in acoustics and the beginning of the modern age of sound.[93] Rayleigh was a product of the Cambridge University mathematical school of the mid-nineteenth century. As senior wrangler in the mathematical tripos of 1865, he was well equipped to handle analytically the problems he encountered in Helmholtz' treatise *Sensations of Tone* and to see that a substantial and authoritative treatise on the whole field of acoustics was needed. His book, which not only brought together material hitherto available only in the journals of learned societies but also presented in detail some of Rayleigh's own contributions, is a work that has long stood as a monument of physical literature. It has had a tremendous influence on the subsequent development of the science of acoustics, particularly on the analytical side.

The Theory of Sound falls into two parts, of which the first relates to mechanical-vibration phenomena of all kinds, including the oscillations of strings, bars, membranes, and plates. (Motions of such structures are, of course, closely connected with the production of sound.) A valuable feature of the treatise is its insistence on the establishment of general principles as well as applications to special cases of practical significance. Rayleigh was an accomplished applied mathematician and developed helpful techniques for the solution of difficult vibration problems. One of these, the basis for what is called the Rayleigh-Ritz method, has had wide modern application, not merely in studying the vibrations of solid structures but also in quantum mechanics. Rayleigh never lost sight of the physical meaning inherent in natural phenomena, and his analysis always has the merit of being applicable in practice.

The second part of *The Theory of Sound* is concerned principally with acoustical propagation through fluid media. In it Rayleigh dealt with such difficult matters as the diffraction of sound waves around obstacles and the general scattering sound undergoes when passing through a medium in which there are many suspended particles, like bubbles in water. (Acoustic diffraction and scattering are much more difficult to handle mathematically than the corresponding phenomena in light because the wave length of ordinary audible sound is of the order of magnitude of the dimensions of the obstacles themselves.) Rayleigh also paid much attention to the geometrical characteristics of the acoustic radiation from vibrating objects (such as pulsating spheres or oscillating discs) that produce "beams" of sound. So thoroughly did Rayleigh treat problems of this kind, and so completely and clearly did he summarize the work of previous investigators on such matters as the attenuation of sound in fluids by various dissipative mechanisms, that when the two volumes of the second enlarged and revised edition of *The Theory of Sound* appeared in 1894 and 1896, many scientists believed that the whole subject of acoustics as a branch of physics was complete, that there was nothing more to learn. These scientists believed that thenceforth acoustics could be dealt with by engineers; this was already happening in the case of the basic theory of electricity and magnetism developed by Ampère, Faraday, and Maxwell, whose scientific efforts laid the foundations of electrical engineering. At that time there was some justification for believing this about acoustics—not because there were no more interesting acoustical phenomena to investigate, but because the experimental means by which these investigations could be carried out practically were not yet available. For example, the work of Rayleigh and his contemporaries strongly suggested that many interesting properties would be found to be associated with sound waves of frequency well above the audible limit, but when *The Theory of Sound* was published, the only practical source of such radiation was the bird whistle.

It seems almost incredible that electromechanical effects were not used earlier as sources of sound of a wide range of frequency. The main difficulty, of course, lay in the inadequacy of the means of producing electrical oscillations and of coupling these oscillations to solid vibrators. The piezo-electric effect, which Pierre and Jacques Curie discovered in 1880, suggested that if there were a way to produce alternating positive and negative electric charges on the opposite faces of a properly cut quartz crystal, the crystal could be made to vibrate. But the successful exploitation of this effect as a source and receiver of sound had to await the invention of the vacuum-tube oscillator and amplifier by Sir John Ambrose Fleming and Lee De Forest in the first two decades of the twentieth century.

The dawn of the twentieth century saw many fundamental problems in acoustics unsolved. On the biological side the nature of hearing was by no means wholly understood, for the relation between the anatomy of the ear and associated nervous system on the one hand and the observed phenomena of audition on the other was by no means clear. Detailed studies of speech were still impractical because there was no adequate way to analyze speech. On the physical side, though a theory existed for the attenuation of sound in fluid media like the atmosphere in terms of the effect of the transport properties (viscosity and heat conduction), it was realized that the results predicted by such effects were not in agreement with experiment. (The predicted values were generally much smaller than the relevant observed values.) The discrepancy was recognized by Lord Rayleigh, who made a shrewd suggestion of a plausible solution of the difficulty in terms of so-called relaxation effects.[94]

Around the turn of the century and for some years afterward, perhaps the biggest obstacle to the further development of acoustics, both as a science and as a branch of technology, was the lack of appropriate sources and receivers of sound. (The latter came to be called transducers.) It was recognized that a great many physical phenomena—mechanical, thermal, electrical, and magnetic—can give rise to sound radiation and conversely can act to transform sound into other physical effects. It was also clear that such transformation, or transduction, could be of enormous practical importance. One obvious example is the telephone, by which sound can be transferred without attenuation over much greater distances than would be possible through the normal sound-transmitting medium itself. This problem was first solved (according to the decision of the courts in the United States after lengthy patent litigation) by Alexander Graham Bell around 1876. The transducing mechanism in this case was, of course, electromagnetic in character. The influence of this invention on the future development of acoustics has been incalculable. In order to improve the telephone huge amounts of money have been expended on research into every aspect of human communication. As a result of the work at the Bell Telephone Laboratories, for example, much more than might ever have been reasonably expected from mere human curiosity is now known about the way the human being hears and speaks. It has also become possible to produce sounds of all frequencies, from a few cycles per second up to several thousand million cycles per second, and to study efficiently the behavior of solid, liquid, and gaseous media exposed to such sounds. At the same time it has become possible to study the interaction between high-frequency radiation—now called ultrasonic—and other physical effects, such as electricity, magnetism, high and low temperatures, and large ranges of pressures.

In the past half century acoustics has shared in the fantastically increasing pace of the progress of science as a whole. To the layman the most obvious developments—like radio and the whole realm of sound recording and reproduction—have been in the technological field rather than in pure science. Less well known, but important for national security, are the transducers designed for transmitting sound underwater in the detection of submarines and other objects. Here the interplay between pure and applied acoustics has been particularly evident. As the power, sensitivity, and efficiency of underwater transducers have increased, greater knowledge of the acoustical properties of sea water has become necessary in order to make full use of the more sophisticated instrumentation. The result has been the opening of a new field of acoustical research as a branch of physical oceanography. Similarly improved instrumentation for the study of speech and hearing has stimulated the creation of wholly new branches of physiology and psychology. The construction of high-power ultrasonic transducers has introduced a new tool for medical research on both the diagnostic and the therapeutic levels. Even music is being influenced by the contemporary ways of producing and amplifying sound.

There are, however, many problems in basic acoustics still to be solved. Among these problems are those that involve the interactions of high-frequency sound radiation with matter in its various phases; but the more purely technological problem of the extension of the practical-frequency limit of ultrasonic radiation must be solved first. The limit in 1967 was about 10^{10} cycles per second. If frequencies of the order of 10^{11} to 10^{14} cycles per second could be realized, our understanding of the nature of the solid, liquid, and gaseous states would be much enhanced. Recent research in which the optical laser has been used as a source of ultrasonic radiation holds out great promise.[95]

Some physicists, carried away by the glamour of high-energy physics and the properties of the solid state, have asserted that the future of a classical field of physics like acoustics lies wholly in its technological applications, that it is played out as physical science. As the history of science in the last half-century shows, there is no more reason to suppose that man will ever run out of questions about acoustics than there is reason to believe that he will run out of questions about the nucleus and its theoretical particles. As investigation proceeds, the boundary lines between the various types of natural phenomena that mankind has artificially erected are becoming less distinct and more unrealistic. The aim of the science of the future is a meaningful synthesis, and acoustics will contribute to that synthesis.

Chapter Five

THE FUTURE OF PHYSICS

It should be clear from the foregoing chapters that the author is convinced that physics will continue to make successful progress through the imaginative construction of theories that allow their inventors great freedom of choice. This freedom implies a large degree of arbitrariness in the way in which the physicist looks at experience, but this very arbitrariness will continue to stimulate the creation of new experience, an open-ended process that seems likely to last as long as the human race continues to retain its curiosity about the way things go and is permitted by circumstances to exercise that curiosity.

The method of physics has been remarkably successful in coping with the aspects of human experience to which it has been applied. This has suggested its application to even wider domains of experience, with already promising results. Geophysics, for example, is producing vast changes in the science of geology: we are learning far more about the nature of the earth's crust as well as the ocean depths through the agency of mechanical radiation of both low and high frequency than would have been thought possible fifty years ago. Astrophysics is now the most important part of astronomy, and through the co-operative efforts of astronomers and physicists, new knowledge of the structure of the universe is being developed. In both geophysics and astrophysics imaginative physicists have led the way to a more profound understanding of our physical environment by suggesting bold and daring theories (among them the theory of the origins of the earth's magnetism and the theory of the expanding universe) and by developing ingenious instruments for performing measurements impossible before the twentieth century.

The situation in the life sciences offers even more exciting possibilities of extending the domain susceptible to physical interpretation. The preoccupation of present-day physiologists and molecular biologists with energy considerations in living systems was foreshadowed by the work of Julius Robert Mayer (1814–78). Mayer, a German physician and amateur physicist, found in the behavior of the living organism a suggestion of the mechanical theory of heat, of which he was one of the principal founders a hundred and twenty-five years ago. If he could return, he would be gratified and amazed to find what a role the concept of energy, the premier concept of physics, now plays in the life sciences. Similarly, Hermann von Helmholtz, a physiologist who became one of the greatest German theoretical physicists of the mid-nineteenth century, initiated a breakthrough in biology and psychology when he decided that all scientists should know more about the way in which the human being sees and hears. Helmholtz' pioneer research on the ear and eye, which used physical methods consistently, laid the groundwork for the elaborate modern investigations in biophysics and psychophysics. Here again precision physical instrumentation is providing tools that Helmholtz and his contemporaries would have longed to have had for the study of the behavior of organisms. However, we shall need thinkers possessing the enormous imaginative theoretical power of Helmholtz before we reach a more thorough understanding of vision and audition. The challenge here to the psychophysicist is very great, but it probably extends even further than to the ordinary modes of sense perception.

Physicists and psychologists tend at present to take a dim view of the phenomena referred to as extrasensory perception and the psychokinetic effect. While it is true that for the most part research in these phenomena has not proved amenable to the standard procedures of physical science, only a dogmatist would eliminate such phenomena from the sphere of physical investigation. There is plenty of evidence, aside from crude card-guessing games, of a subtle environmental influence on sensitive organisms that exists outside the range normally associated with the conventional avenues of sense perception. The exploration in depth of these phenomena will undoubtedly form an important occupation for psychophysicists of the future. To appreciate the possibilities inherent in this area, one has only to think of the peculiar nervous effects produced in some individuals by inaudible high-intensity, high-frequency sound and the even more obvious influence of atmospheric temperature, pressure, and humidity on persons suffering from certain types of diseases.

Broadening the domain of experience susceptible to physical investigation and interpretation involves both positive benefits and possible dangers. On the positive side it marks definite progress toward the unity of science that was the goal of the early philosophers, whose grasp of it was of course severely limited by the very limitation of their experience. The method of physics may well prove to be a kind of philosopher's stone, bringing an understanding of all human experience under the domination of a few simple but comprehensive constructs, like the construct of energy. In the light of the intense specialization and fragmentation of much modern scientific endeavor, the attainment of this ideal in modern science may seem unrealistic to many. But it must be emphasized that this very specialization, largely made possible by physical ideas and instrumentation, stimulates the quest for unifying concepts that physics, and probably physics alone, can provide. It is reasonable to foresee that the time is not too far distant when the specific sciences now professionally recognized will disappear, to be merged into a single discipline dedicated to the understanding of all human experience.

There is still another positive result of the widening of the range of physical experience. This is a mitigation of the possible harm inherent in the tendency of some physicists to assume rather dogmatically that the future development of "real" physics, as they prefer to call it, lies wholly in one direction—as when certain nuclear physicists, carried away by their enthusiasm for the fascinating but often mysterious results obtained by hurling atomic projectiles at ever-increasing velocity at the nuclei of atoms, have felt that the physics of the future is concerned almost entirely with the elementary particles that result from these atomic barrages. Their enthusiasm is understandable, but the dogmatism that emerges from it could prove embarrassing in terms of the distribution of the support of research. Modern physical research is very expensive, and it would be unfortunate if the zeal of one school of thought should lead to a disproportionate expenditure of money for its research and the deprivation of other lines that might conceivably lead to results of equal interest in the long run. However, a very wide diversification of interest results from the many avenues along which physics is moving in its interaction with other sciences; this is useful insurance against an eventual bias. Perhaps those scientists who deplore certain aspects of space technology are justified in doing so; however, space technology has stimulated the attention of physicists to a variety of problems that are unquestionably enlarging human experience. Diversification is healthy.

There is, to be sure, a negative side to the extension of the domain of experience that is coming under the scrutiny of physicists. As research becomes more specialized and expensive, there is a tendency to replace individual investigation by team research and to develop a hierarchy of responsibility in research institutions. This worked well during the stress of war, when research was directed toward very practical problems of crucial importance for national survival, and even in peacetime it has its merits in applied research and development. But in the basic research of the future this tendency may inhibit the individual thinker and the ingenious, unconventional worker; it must not be forgotten that the important fundamental concepts come from the minds of individuals. Let us hope that in the future, as in the past, there will be intuitive individuals who have enough toughmindedness and independence of spirit to keep them from being swallowed up by any highly organized system of science.

The psychology of physics—the problem of how physicists select the kinds of experience they do for special investigation and how they arrive at the methods they use in building theories to explain experience—was mentioned in Chapter One, but no attempt was made to cope with it. We do not understand how physics is created and can but analyze logically the results of the creation. Although the attempts that have thus far been made to investigate the psychology of physical invention have merely scratched the surface of the subject, it is clear that this is a field that will receive much attention in the future, and the combined efforts of physicists and psychologists may well have a significant effect on the kind of physics our descendants will be faced with. Among the questions that studies of this sort might tackle is that of whether future physicists will be able to provide a more adequate grasp of experience than before by abandoning the spatio-temporal matrix, hitherto the basis of all physical theories, in favor of other ways of categorizing experience. This may seem fanciful, but it has already been suggested by some elementary particle physicists. Along the same line would be the abandonment of the object-instrument dichotomy in physical measurement that was mentioned in Chapter Three.

One of the most serious problems faced by the physicist—and, indeed, of all scientists—of the future is that posed by communication, as anyone who has heard of modern information theory and cybernetics is aware. Communication impinges on physics at many levels. Indeed the origin of physics itself may be regarded as the communication of knowledge to the perceptive human individual by something variously called Nature, the external world, or simply the world of our experience. In this connection communication means the whole process involving the stimuli we receive as organisms coupled with the responses we make to these stimuli. From this point of view communication is the basic element in all science; everything that takes place in any physical laboratory or other scientific laboratory can be analyzed in terms of communication.

Communication enters into physics on another level, for in building physics as a science the physicist has to go through the process of communicating to himself (literally talking to himself about his ideas and constructions) and then of communicating his ideas to others who are presumably equally interested in the same forms of experience. In so far as physics is a social activity, it can flourish only through the interaction of physicists, and this can take place only by means of communication. Finally, if a science like physics is to be tolerated and supported by the general community, the general public must have some idea of what physicists are doing. This again involves communication, which may take the form of written and oral presentation either in popular language or in the language of technological application resulting directly from basic physical research.

As the domain of physical experience has widened, the language of physics, mainly mathematical in character, has grown more elaborate and extensive. This has made the problem of communication more difficult—from the standpoint of learning the language and from the sheer physical exertion of coping with the enormous amount of material published each year in thousands of journals throughout the world. The very difficulty of locating material of interest to a particular investigator—to say nothing of the problems involved in reading and understanding it—has made it necessary to initiate research projects directed toward developing more effective retrieval methods. The researcher wants to be able to locate as quickly as possible all the literature that has appeared in print on a given topic during a certain period of time; using the ordinary journal index, however, even if it has been carefully prepared, is very slow. The ultimate solution will undoubtedly be to use high-speed digital computers with sufficient storage capacity and memory to handle not merely titles and authors of research articles but also the whole text of the articles themselves and with the ability to reproduce or demand any desired amount of relevant material. This solution will make the present-day archival research library obsolete.

Because communication with computers and even with human beings is most conveniently carried out through the agency of a language, any intensive study of communication demands a better understanding of the nature of language. Physics itself is contributing to the study of communication through the theory of thermodynamics, one of the grandest of all physical theories. The result may well be not only the invention of new kinds of mathematics—mathematics itself being a very powerful language —but also the more efficient use of the language of ordinary speech, whose redundancy may be of value in ordinary human intercourse but is highly uneconomical for scientific purposes, particularly in translating scientific material from one language to another.

No one is in a position to make a valid projection over a wide span of time of the future of a science like physics with any greater likelihood of success than the predictors of the course of human history in general. We can be sure, however, that as long as human curiosity and imaginative power endure, physics will always make a significant contribution to the description, creation, and understanding of nature.

NOTES

NOTES TO CHAPTERS TWO, THREE, AND FOUR

Chapter Two

1. *An Essay on the Psychology of Invention in the Mathematical Field* (Princeton: Princeton University Press, 1945).
2. Moles, *La Création scientifique* (Geneva: Editions René Kister, 1957); Leclercq, *Traité de la méthode scientifique* (Paris: Dunod, 1964).
3. *The Nature of Physical Reality* (New York: McGraw-Hill Book Co., 1950), pp. 69 ff.
4. *Ibid.*, pp. 232 ff.; F. S. C. Northrop, *The Meeting of East and West* (New York: The Macmillan Co., 1946), p. 443.
5. Earle H. Kennard, *Kinetic Theory of Gases* (New York: McGraw-Hill Book Co., 1938), pp. 180 ff.
6. Max Born, *Atomic Physics,* trans. John Dougall (New York: G. E. Stechert and Co., 1935), p. 192.
7. *Ibid.*, p. 189.
8. *Ibid.*, pp. 184 ff.
9. Antoine L. Lavoisier, *Traité élémentaire de chimie* (Paris: Cuchet, 1789).
10. "An Inquiry concerning the Source of the Heat Which Is Excited by Friction," *Philosophical Transactions of the Royal Society,* LXXXVIII (1798), 80.
11. This matter is discussed in some detail in Robert Bruce Lindsay, *The Role of Science in Civilization* (New York: Harper & Row, 1963).
12. *Dialogues concerning Two New Sciences,* trans. Henry Crew and Alfonso DeSalvio (New York: The Macmillan Co., 1914), p. 160.
13. Henry Cavendish, "Experiments To Determine the Density of the Earth," *Philosophical Transactions of the Royal Society,* LXXXVIII (1798), 469.
14. Robert Bruce Lindsay and Henry Margenau, *Foundations of Physics* (New York: Dover Publications, Inc., 1957), pp. 128 ff.
15. *The Aim and Structure of Physical Theory,* trans. Philip P. Wiener (Princeton: Princeton University Press, 1954), pp. 188 ff.
16. *Mathematical Principles of Natural Philosophy,* trans. Florian Cajori (Berkeley: University of California Press, 1934), p. 398.

Chapter Three

1. For a somewhat more detailed discussion of a scientist's view of the relation between science and philosophy, see Robert Bruce Lindsay, *The Role of Science in Civilization* (New York: Harper & Row, 1963), chap. iv.
2. Robert Bruce Lindsay and Henry Margenau, *Foundations of Physics* (New York: Dover Publications, Inc., 1957), pp. 65 ff.

3. *Ibid.*, pp. 68 ff.

4. *Mathematical Principles of Natural Philosophy,* trans. Florian Cajori (Berkeley: University of California Press, 1934), p. 6.

5. "The Logical Structure of Physics," *Synthèse,* XIV (1962), 110; "The Conceptual Structure of Physics," *Reviews of Modern Physics,* XXXV (1963), 151.

6. See, for example, Robert Bruce Lindsay, *Introduction to Physical Statistics* (New York: J. Wiley & Sons, Inc., 1941), p. 109.

7. *Elementary Principles in Statistical Mechanics* (New Haven: Yale University Press, 1903); see also Lindsay, *Introduction to Physical Statistics,* pp. 102 ff.

8. These are clearly set forth in G. J. Whitrow, *The Natural Philosophy of Time* (London and Edinburgh: Thomas Nelson and Sons, Ltd., 1961), pp. 176 ff.

9. Herbert Dingle, "Relativity and Space Travel," *Nature,* CLXXVII (1956), 782.

10. See, for example, W. M. McCrea's untitled reply to Dingle, *ibid.,* p. 784.

11. Whitrow, *Natural Philosophy,* pp. 219 ff.

12. *The Logic of Modern Physics* (New York: The Macmillan Co., 1927).

13. Robert Bruce Lindsay, "A Critique of Operationalism in Physics," *Philosophy of Science,* IV (1937), 456.

14. "Operational Analysis," *Philosophy of Science,* V (1938), 114.

15. "The Nature of Some of Our Physical Concepts," *British Journal for the Philosophy of Science,* I (1950/51), 257, and II (1951/52), 25 and 142.

16. *Ibid.,* I, 260.

17. *Advancement of Learning* in Vol. XXI, Part I of *A Library of Universal Literature,* ed. Joseph Devey (New York: P. F. Collier and Son, 1901), pp. 17, 30.

18. See, for example, Lindsay and Margenau, *Foundations,* p. 417.

19. See, for example, David Bohm, *Causality and Chance in Modern Physics* (Princeton: D. Van Nostrand Co., Inc., 1957); also the Proceedings of the Ninth Symposium of the Colston Research Society, *Observation and Interpretation: A Symposium of Philosophers and Physicists,* ed. S. Körner (London: Butterworths Scientific Publications, 1957).

20. See, for example, Lindsay and Margenau, *Foundations,* pp. 182 ff.

21. *Relativity Theory of Protons and Electrons* (Cambridge: Cambridge University Press, 1936), p. 319.

22. "Ziele und Wege der physikalischen Erkenntnis," *Allgemeine Grundlagen der Physik,* ed. H. Thirring, in *Handbuch der Physik,* ed. H. Geiger and K. Scheel, IV (Berlin: Julius Springer, 1929), 1–80.

23. Lindsay and Margenau, *Foundations,* pp. 79 ff.

24. *Ibid.,* p. 91.

25. *The Scientific Papers of James Clerk Maxwell,* ed. W. D. Niven (Cambridge: Cambridge University Press, 1890), II, 219 ff.

26. For detailed discussion of Russell's definition see Lindsay, *Role of Science,* pp. 186 ff.

27. *A Mathematician's Apology* (Cambridge: Cambridge University Press, 1940).

Chapter Four

1. See, for example, the brief review in Robert Bruce Lindsay, *The Role of Science in Civilization* (New York: Harper & Row, 1963), chap. v.

2. *Philosophy of History, An Introduction* (New York: Harper Torchbooks, Harper & Brothers, 1960), p. 29.

3. This has been done, of course, most ingeniously and convincingly by Otto Neugebauer, *The Exact Sciences in Antiquity* (2d ed.; Providence: Brown University Press, 1957).

4. Truesdell, *The Rational Mechanics of Flexible or Elastic Bodies, 1638–1788: Introduction to Leonhardi Euleri Opera omnia*, Ser. 2, XI (Zurich: Orell Füssli, 1960). The first of eight projected volumes of Newton's publications has recently appeared: *The Mathematical Papers of Isaac Newton, Volume I: 1664–66*, ed. D. T. Whiteside (Cambridge and New York: Cambridge University Press, 1967); also four volumes of *The Correspondence of Isaac Newton*, ed. H. W. Turnbull and J. F. Scott (Cambridge and New York: Cambridge University Press, 1959–67), have been published, and three additional volumes are in preparation.

5. See Goudsmit's interesting article on the discovery of electron spin, "Pauli and Nuclear Spin," *Physics Today*, XIV, No. 6 (June, 1961), 18.

6. "History in the Education of Scientists," *American Scientist*, XLVIII (1960), 528.

7. See Lindsay, *Role of Science*, pp. 120 ff.

8. *Bibliographical History of Electricity and Magnetism* (London: C. Griffin & Co., Ltd., 1922), p. 46.

9. *William Gilbert of Colchester, Physician of London, On the Loadstone and Magnetic Bodies, and on the Great Magnet of the Earth*, trans. P. F. Mottelay (New York: J. Wiley & Sons, 1893). This translation is cited as the most accessible one. Reference should also be made here to the later translation brought out in 1900 in London for the Gilbert Club by the Chiswick Press. Professor S. P. Thompson was the chief editor. He also prepared an elaborate set of notes in connection with the translation. Only 250 copies of the edition were printed, and it is already very scarce. The Gilbert Club was founded in 1889, largely through the enthusiastic interest of Thompson, who was preparing a translation of *De magnete* for the tercentenary celebration in 1900. The translation is a page-for-page reproduction of the original edition in every respect but the language. It is now generally agreed that the Gilbert Club translation is the definitive one. Over a period of ten years, Thompson brought out a number of privately printed articles and pamphlets concerning Gilbert. For a description and discussion of these see J. S. and H. G. Thompson, *Silvanus Phillips Thompson* (New York: Dutton, 1920).

10. *William Gilbert*, p. xlvii.

11. *The History of the Worthies of England* (London: F. C. and J. Rivington, 1811), I, 352.

12. *Dialogues concerning the Two Chief World Systems: Ptolemaic and Copernican,* trans. Stillman Drake (Berkeley and Los Angeles: University of California Press, 1955), p. 406.

13. Francis Bacon, *The Novum Organon or a True Guide to the Interpretation of Nature,* trans. G. W. Kitchin (Oxford: Oxford University Press, 1855), pp. 26, 27.

14. See, for example, the account by John Robinson in his *System of Mechanical Philosophy* (London, 1822).

15. William Cecil Dampier-Whetham, *A History of Science and Its Relations with Philosophy and Religion* (New York: The Macmillan Co., 1930), p. 136.

16. *William Gilbert,* pp. 294, 298; see also Robert Bruce Lindsay, "William Gilbert and Magnetism in 1600," *American Journal of Physics,* VIII (1940), 271–82.

17. *William Gilbert,* p. 292.

18. *Worthies of England,* p. 352.

19. *Areopagitica,* in *Complete Prose Works of John Milton,* II, ed. Ernest Sirluck (New Haven: Yale University Press, 1959), 537–38.

20. Alexandre Koyré, "A l'Aube de la science classique," Part I of *Etudes galiléennes,* in *Actualités scientifiques et industrielles,* No. 852 (Paris: Hermann & Co., 1939), p. 47.

21. *Le Opere di Galileo Galilei,* X, ed. Giorgio Abetti (Florence: G. Barbèra, 1934), 115.

22. *Dialogues concerning Two New Sciences,* trans. Henry Crew and Alfonso DeSalvio (New York: The Macmillan Co., 1914), p. 167.

23. William Whewell, *History of the Inductive Sciences* (3d ed.; New York: D. Appleton and Co., 1874), p. 24.

24. Morris R. Cohen and Israel E. Drabkin, *A Source Book in Greek Science* (New York: McGraw-Hill Book Co., 1948), pp. 286–310.

25. *Two New Sciences,* pp. 95–108.

26. In Truesdell, *Rational Mechanics,* a mine of detailed information on the history of acoustics in the seventeenth and eighteenth centuries, see especially pp. 28–37 for developments in the theory of vibration.

27. Robert T. Gunther, *The Life and Work of Robert Hooke,* in *Early Science in Oxford,* VI (Oxford: Oxford University Press, 1930), 57.

28. "Système général des intervalles des sons," *Histoire de l'Académie royale des sciences* (Paris), vol. for 1701, pp. 297–364.

29. "De motu nervi tensi," *Philosophical Transactions of the Royal Society,* XXVIII (1713), 11–21.

30. *Rational Mechanics,* p. 130.

31. "Reflexions et éclaircissements sur les nouvelles vibrations des cordes, exposées dans les Mémoires de l'Académie de 1747 et 1748," *Mémoires de l'Académie royale des sciences et belles lettres* (Berlin), vol. for 1755, p. 147.

32. "Recherches sur la courbe que forme une corde tendue mise en vibration,"

Mémoires de l'Académie royale des sciences et belles lettres (Berlin), vol. for 1747, p. 214.

33. The most complete study of the contributions of Euler to acoustics and the relation of his work to that of his contemporaries is that of Clifford Truesdell in *Rational Mechanics*.

34. "Dr. Wallis's Letter to the Publisher concerning a New Musical Discovery, Written from Oxford, March 19, 1676," *Philosophical Transactions of the Royal Society,* XIII (1677), 839.

35. See n. 28.

36. See n. 31.

37. "Remarques sur les mémoires précédens de M. Bernoulli," *Mémoires de l'Académie des sciences et belles lettres* (Berlin), vol. for 1753, p. 196.

38. *La Théorie analytique de la chaleur* (Paris: F. Didot, 1822).

39. "Recherches sur la nature et la propagation du son," *Oeuvres de Lagrange,* I, ed. J. A. Serrett (Paris: Gauthier-Villars, 1867), 39 ff.

40. "De motu corporum flexibilium," *Leonhardi Euleri Opera omnia,* Ser. 2, X (Zurich: Orell Füssli, 1947), 177 ff.

41. *Rational Mechanics,* pp. 263–73.

42. "Dissertatio physica de sono," *Leonhardi Euleri Opera omnia,* Ser. 3, I (Leipzig: Teubner, 1926), 182.

43. "Eclaircissements plus détaillés sur la génération et la propagation du son, et sur la formation de l'écho," *Mémoires de l'Académie des sciences et belles lettres* (Berlin), vol. for 1765, pp. 335–63; *Leonhardi Euleri Opera omnia,* Ser. 3, I, 540.

44. Gunther, *Robert Hooke,* in *Early Science,* VIII, 119–52.

45. Truesdell, *Rational Mechanics,* p. 165.

46. *Entdeckungen über die Theorie des Klanges* (Leipzig: Weidmann Erben und Reich, 1787).

47. "Uber das Gleichgewicht und die Bewegung einer elastischen Scheibe," *Crelle's Journal,* XL (1850), 51.

48. "Mémoire sur l'équilibre et le mouvement des corps élastiques," *Mémoires de l'Académie royale des sciences* (Paris), VIII (1829), 357–570.

49. *Theorie der Elasticität fester Körper* (Leipzig: Teubner, 1862).

50. "De motu vibratorio tympanorum," *Novi Commentarii Academiae Scientarum Imperialis Petropolitanae,* X (1764), 243; *Rational Mechanics,* pp. 330–34.

51. For an excellent review of the history of electroacoustics see Frederick V. Hunt, *Electroacoustics* (Cambridge: Harvard University Press, 1954).

52. Pierre and Jacques Curie, "Cristallophysique," *Comptes rendus de l'Académie des sciences* (Paris), XCI (1880), 294; "Physique Cristallographique," *ibid.,* 383.

53. Osborne Reynolds, *Memoir of James Prescott Joule* (Manchester, Eng.: Literary and Philosophical Society, 1892), p. 101.

54. Helmholtz' results were set forth in a treatise that is one of the great master-pieces of acoustics, *Die Lehre von den Tonempfindungen als physiologische Grundlage für die Theorie der Musik* (Brunswick: F. Vieweg, 1863); see also the English translation from the third German edition, *Sensations of Tone,* trans. A. J. Ellis (London: Longmans, Green, 1885); also Sir Richard Paget, *Vowel Resonances* (London: International Phonetic Association, 1922).

55. Cohen and Drabkin, *Source Book,* pp. 288 ff.

56. *Ibid.,* pp. 307 ff.

57. See Robert Bruce Lindsay, "Pierre Gassendi and the Revival of Atomism in the Renaissance," *American Journal of Physics,* XIII (1945), 235.

58. *New Experiments Physico-Mechanical Touching the Air* (2d ed.; Oxford: T. Robinson, 1662), Experiment 27, p. 103; see also p. 143.

59. J. M. A. Lenihan, "Mersenne and Gassendi, an Early Chapter in the History of Sound," *Acustica,* II (1951), 96.

60. Dayton C. Miller, *Anecdotal History of the Science of Sound* (New York: The Macmillan Co., 1935), p. 20.

61. "Experimenta et observationes de soni motu alliisque ad id attinentibus," *Philosophical Transactions of the Royal Society,* XXVI (1708), 1.

62. E. Cherbuliez, "Geschichtliche Übersicht der Untersuchungen über die Schall-fortpflanzungsgeschwindigkeit in der Luft," *Mitteilungen naturforschenden Gesellschaft* (Bern), vol. for 1870, pp. 141–91; vol. for 1871, pp. 1–28; see also J. M. A. Lenihan, "The Velocity of Sound in Air," *Acustica,* II (1952), 205.

63. *American Institute of Physics Handbook* (2d ed.; New York: McGraw-Hill Book Co., 1963), pp. 3–64, 65.

64. *Annales de chimie et de physique,* XIII (1808), 5.

65. Colladon and Sturm, "Sur la Compression des liquides," *Annales de chimie et de physique,* XXXVI (1827), 113, 225.

66. *Souvenirs et mémoires: Autobiographie de J-Daniel Colladon* (Geneva, 1893).

67. "Sur la vitesse du son dans l'air et dans l'eau," *Annales de chimie et de physique,* III (1816), 238.

68. "Uber eine neue Art akustischer . . . ," *Poggendorfs Annalen der Physik,* CXXVII (1866), 497.

69. "Sur l'intégration de quelques équations linéaires aux différences partielles . . . ," *Mémoires de l'Académie des sciences* (Paris), vol. for 1818, p. 121.

70. "Sur le mouvement des fluides élastiques dans des tuyaux cylindriques . . . ," *Mémoires de l'Académie des sciences* (Paris), vol. for 1817, p. 305.

71. Green, "On the Reflexion and Refraction of Sound," *Proceedings and Transactions of the Cambridge Philosophical Society,* VI (1838), 403; for Poisson, see n. 48.

72. "De la propagation du son," *Mémoires de l'Académie des sciences et belles lettres* (Berlin), vol. for 1759, p. 185; or *Leonhardi Euleri Opera omnia,* Ser. 3, I, 428.

73. Riemann, "Uber die Fortpflanzung ebener Luftwellen von endlichen Schwingungsweite," *Abhandlungen der Königlichen Gesellschaft der Wissenschaften zu Göttingen,* VIII (1860), 43; and Earnshaw, "On the Mathematical

Theory of Sound," *Philosophical Transactions of the Royal Society*, CL (1860), 133.

74. "Mémoire sur la théorie du son," *Journal de l'école polytechnique*, VII (1808), 319.

75. For detailed bibliographical references see Richard Courant and Kurt Otto Friedrichs, *Supersonic Flow and Shock Waves* (New York: Interscience Publishers, 1948), pp. 438 ff.

76. "Uber die Empfindlichkeit des Gehörorgans," *Poggendorfs Annalen der Physik*, XX (1830), 290.

77. See, for example, Ludwig F. W. A. Seebeck, "Beobachtungen über einige Bedingungen der Entstehung von Tönen," *Poggendorfs Annalen der Physik*, LIII (1841), 417; Helmholtz, *Sensations of Tone*, pp. 11, 13, 162, 182, 372; Koenig, "Uber die höchsten hörbaren und unhörbaren Töne . . . ," *Annalen der Physik*, CCCV (1899), 626, 721.

78. *Quelques expériences d'acoustique* (Paris: A. Lahure, 1882).

79. Toepler and Boltzmann, "Uber eine neue optische Methode . . . ," *Poggendorfs Annalen der Physik*, CXLI (1870), 321.

80. "Uber die Definition des Tones . . . ," *Poggendorfs Annalen der Physik*, LIX (1843), 497.

81. See n. 54.

82. Upham, "A Consideration of Some of the Phenomena and Laws of Sound and Their Application in the Construction of Buildings Designed Especially for Musical Effects," *American Journal of Science and Arts*, Ser. 1, LXV (1853), 215, 348; LXVI (1853), 21; Henry, "Acoustics Applied to Public Buildings," *Annual Report of the Smithsonian Institution* (Washington, D.C., 1856), 221.

83. *Collected Papers on Acoustics* (Cambridge: Harvard University Press, 1927; New York: Dover Publications, Inc., 1964).

84. Gunther, *Robert Hooke*, in *Early Science*, VI, 330 ff.

85. A. Laboulbène, "Biographies scientifiques: Laënnec," *Revue scientifique*, Ser. 4, IX (1898), 40.

86. *The Scientific Papers of Sir Charles Wheatstone* (London: Physical Society of London, 1879).

87. "On the Influence of Musical Sounds on the Flame of a Jet of Coal Gas," *Philosophical Magazine*, XV (1858), 235.

88. "On Sounding and Sensitive Flames," *Philosophical Magazine*, XXXIII (1867), 92; "On the Action of Sonorous Vibrations on Gaseous and Liquid Jets," *ibid.*, 375; also by Tyndall, *Sound* (New York: D. Appleton and Co., 1876), chap. vi.

89. "Die manometrischen Flammen," *Poggendorfs Annalen der Physik*, CXLVI (1872), 161.

90. "Inscription automatique des sons de l'air au moyen d'une oreille artificielle," *Comptes rendues de l'Académie royale des sciences* (Paris), LIII (1861), 108.

91. "A Method of Recording Articulate Vibrations by Means of Photography," *American Journal of Science and Arts*, Ser. 3, XVI (1878), 54.

92. *The Science of Musical Sounds* (New York: The Macmillan Co., 1916), pp. 78 ff.

93. The second edition (Vol. I, 1894; Vol. II, 1896) has been reprinted with a historical introduction by Robert Bruce Lindsay (New York: Dover Publications, 1945).

94. Lord Rayleigh, "On the Cooling of Air by Radiation and Conduction, and on the Propagation of Sound," *Philosophical Magazine,* XLVII (1899), 308.

95. Details can be found in the *Journal of the Acoustical Society of America.* See in particular "Symposium on Unsolved Problems in Acoustics," ed. Robert Bruce Lindsay, XXX (1958), 275; also "Report of the Acoustics Research Survey Committee," Richard K. Cook, chmn., XXXVII (1965), 392.

INDEX

INDEX